CITIZEN SCIENCE

CITIZEN SCIENCE

Public Participation in Environmental Research

EDITED BY

JANIS L. DICKINSON AND RICK BONNEY

FOREWORD BY

RICHARD LOUV

AFTERWORD BY

JOHN W. FITZPATRICK

COMSTOCK PUBLISHING ASSOCIATES

A DIVISION OF CORNELL UNIVERSITY PRESS

ITHACA AND LONDON

First published 2012 by Cornell University Press
Printed in the United States of America

Library of Congress Cataloging-in-Publication Data

Citizen science: public participation in environmental research / edited by Janis L. Dickinson and Rick Bonney; foreword by Richard Louv; afterword by John W. Fitzpatrick.
      p. cm.
  Includes bibliographical references and index.
  ISBN 978-0-8014-4911-6 (cloth: alk. paper)
  1. Environmental sciences—Research—Citizen participation.
I. Dickinson, Janis L., 1955– II. Bonney, Rick, 1954–
  GE70.C543 2012
  304.2072—dc23
                        2011037482

Cloth printing   10 9 8 7 6 5 4 3 2 1

Contents

Foreword

The original use of "amateur" probably came from the French form of the Latin root, *amator*: lover, or lover of, or devotee. In Thomas Jefferson's agrarian time, few people made their living as scientists; most were amateurs, as was Jefferson. As an amateur naturalist, he tutored Meriwether Lewis in the White House before sending him off to record the flora and fauna of the West. Since then, the word "amateur" has become something of a pejorative. But now the stage is set for the return of the amateur, in a twenty-first-century incarnation, as the citizen scientist. The popular spirit is willing. Citizen scientists already comprise the vanguard of a new nature movement, one as focused on human restoration as it is on the restoration and expansion of natural habitats. Partly guided by leaders with impressive academic credentials, this is largely a revolution of amateurs. Citizen scientists/naturalists are young and old; they're teachers, journalists, and plumbers. In my home bioregion, they track wild boars and chart bird migration routes. They sit on mountaintops for weeks in the Anza-Borrego Desert to record the ghostly presence of bighorn sheep and help trail mountain lions; working with a university marine biologist, they tag and follow threatened shark species.

As Janis Dickinson and Rick Bonney describe in the volume you now hold, the citizen science movement is nurtured by an array of organizations. Project BudBurst, a campaign with nationwide reach, enlists volunteers to mark the first flush of spring growth, as well as the initial flowers and ripe fruit among a wide range of plant species, native and exotic. The Cornell Lab of Ornithology's Project FeederWatch has for many years enlisted the interest and sharp eyes of amateur birders to help scientists understand movements of winter bird populations. In addition, the National Wildlife

Federation, with over four million members, is expanding its efforts to recruit young people to become NWF–certified citizen naturalists with significant emphasis on participation in a variety of citizen science projects. And the Woodland Trust, the UK's leading woodland conservation charity, is expanding its network of thousands of registered nature "recorders." This is only a sampling of the organizations nurturing this movement.

Increasingly, our sense of kinship with other species depends on the fortitude of citizen scientists and the support that they receive. Here's one example of why citizen scientists are so important: In my county, 400 volunteers helped compile the landmark San Diego County Bird Atlas not long before massive firestorms burned 20 percent of the county's land surface, possibly wiping out entire bird populations. Those volunteers expanded our knowledge of the nearly 500 species of birds that live or pass through this bioregion. As the saying goes, we don't know what we've got until its gone—and we don't know what can be restored if we never recognized its existence in the first place.

Citizen scientists collect more than data. They gather meaning. The citizen science movement deserves more public attention and encouragement. What has been missing up to now is an accurate account of the history of citizen participation in environmental science, as well as a clear-eyed appraisal of current trends and an assessment of how best to approach the future through careful project design and the skillful management of information technology. In other words, we have needed *this book*. Thanks to the efforts of Dickinson and Bonney, students and researchers, organizers and volunteers now have a resource that provides best practice descriptions, Internet links to ongoing projects, and inspiration. More than ever, we need solid environmental research and the tools to make sense of all the data. We need greater scientific literacy and the rapid communication of study results. In response to concern about the shortage of professional naturalists and taxonomists, citizen scientists will play an increasingly important—a crucial—role in the expansion of that knowledge. This book challenges professional scientists to find new and better ways to engage an interconnected public with new electronic tools for observing, recording, and reporting.

In recent decades, we have witnessed a disturbing disconnection of children and the young from the natural world. An expansion of the citizen scientist movement, particularly if it reaches out directly to recruit the young, will be a powerful antidote to what I've called the nature-deficit disorder of our species. This book will help make the life on our shared planet more visible for everyone; it will help increase the number of front-line citizen scientists, who, in cities and the countryside and the wilderness, serve the kinship of all species. Here is an essential tool for Thomas Jefferson's spiritual heirs.

RICHARD LOUV, author of *The Nature Principle: Human Restoration and the End of Nature-Deficit Disorder* and *Last Child in the Woods*

Notes on Contributors

Paul Alaback is professor emeritus of ecology at the University of Montana and lead science adviser for Project BudBurst.

Rick Bonney is senior extension associate and director of program development and evaluation at the Cornell Lab of Ornithology.

David N. Bonter is assistant director of citizen science and project leader for FeederWatch at the Cornell Lab of Ornithology.

Wesley R. Brooks is a PhD candidate in the Ecology and Evolution Graduate Program at Rutgers, The State University of New Jersey.

Miyoko Chu is senior director of communications at the Cornell Lab of Ornithology.

Caren B. Cooper is a research associate in bird population studies at the Cornell Lab of Ornithology.

André A. Dhondt is Edwin H. Morgens Professor of Ornithology in the Department of Ecology and Evolutionary Biology, Cornell University, and director of bird population studies at the Cornell Lab of Ornithology.

Janis L. Dickinson is professor of natural resources at Cornell University and Arthur A. Allen Director of Citizen Science at the Cornell Lab of Ornithology.

Joan G. Ehrenfeld is professor of ecology, evolution, and natural resources at Rutgers, The State University of New Jersey.

Jennifer Fee is manager of K–12 programs at the Cornell Lab of Ornithology.

Daniel Fink is a research associate and statistician in information science at the Cornell Lab of Ornithology.

John W. Fitzpatrick is professor of ecology and evolutionary biology at Cornell University and executive director of the Cornell Lab of Ornithology.

Cecilia Garibay is principal of Garibay Group.

Steven A. Gray is assistant professor of natural resources and environmental management at the University of Hawaiʻi at Mānoa.

Jeremy J.D. Greenwood is a research fellow (formerly director) of the British Trust for Ornithology and honorary professor in the Centre for Ecological and Environmental Monitoring, University of St. Andrews, Scotland.

Ralph S. Hames is a research associate in conservation science at the Cornell Lab of Ornithology.

Kayri Havens is director of plant science and conservation at the Chicago Botanic Garden and lead botanist for Project BudBurst.

Sandra Henderson is director of Project BudBurst, Education and Public Engagement, National Ecological Observatory Network (NEON).

Cindy E. Hmelo-Silver is professor of educational psychology at the Rutgers Graduate School of Education and coeditor in chief of the *Journal of the Learning Sciences*.

Wesley M. Hochachka is a senior research associate and assistant director of Bird Population Studies at the Cornell Lab of Ornithology.

David V. Howe is an educational technologist in the Department of Ecology, Evolution, and Natural Resources at Rutgers, The State University of New Jersey.

Rebecca C. Jordan is associate professor of environmental education and citizen science and director of the Program in Science Learning at Rutgers, The State University of New Jersey.

Steve Kelling is director of information science at the Cornell Lab of Ornithology.

Barbara A. Knuth is vice provost and dean of the Graduate School and professor of natural resource policy and management at Cornell University.

Marianne E. Krasny is professor of natural resources and chair of the Department of Natural Resources at Cornell University.

Kristi S. Lekies is an assistant professor in the School of Environment and Natural Resources at The Ohio State University.

Patricia Leonard is a staff writer and Great Backyard Bird Count coordinator at the Cornell Lab of Ornithology.

James D. Lowe is a research biologist and project leader for Birds in Forested Landscapes at the Cornell Lab of Ornithology.

Peter P. Marra is a research scientist at the Smithsonian Conservation Biology Institute's Migratory Bird Center.

Kevin McGarigal is associate professor of natural resources conservation at the University of Massachusetts, Amherst.

Kirsten K. Meymaris is a Web technologist and instructional designer for Project BudBurst.

Karen S. Oberhauser is associate professor of fisheries, wildlife, and conservation biology at the University of Minnesota.

Tina Phillips is manager of program development and evaluation at the Cornell Lab of Ornithology.

Karen Purcell is project leader for Celebrate Urban Birds at the Cornell Lab of Ornithology.

Robert Reitsma is a research technician at the Smithsonian Conservation Biology Institute's Migratory Bird Center.

Kenneth V. Rosenberg is a senior research associate and director of conservation science at the Cornell Lab of Ornithology.

Jennifer L. Shirk is a PhD candidate in the Department of Natural Resources at Cornell University and project leader for the Citizen Science Toolkit at the Cornell Lab of Ornithology.

Flisa Stevenson is the National Wildlife Refuge System Visitor Services Program liaison at the Cornell Lab of Ornithology.

Keith G. Tidball is a senior extension associate in the Department of Natural Resources and associate director of the Civic Ecology Lab at Cornell University.

Nancy M. Trautmann is a senior extension associate in the Department of Natural Resources at Cornell University and director of education at the Cornell Lab of Ornithology.

Heather A. Triezenberg is program director for social science at the National Oceanic and Atmospheric Administration's National Sea Grant College Program.

Dennis L. Ward is lead Web technologist for Project BudBurst, Education and Public Engagement, National Ecological Observatory Network (NEON).

Nancy M. Wells is associate professor of design and environmental analysis at Cornell University.

Y. Connie Yuan is assistant professor of communication at Cornell University.

Benjamin Zuckerberg is assistant professor of forest and wildlife ecology at the University of Wisconsin–Madison.

Acknowledgments

We are grateful to our editor, Heidi Lovette, for seeing us through the process of creating this book. Indeed, the book was her inspiration in the first place, and her advice was invaluable as the book took shape. We also express gratitude to Katherine Hue-tsung Liu and Susan P. Specter for shepherding the manuscript through the process of publication; our copy editor, Marie Flaherty-Jones, for fixing what was wrong; the editorial board of Cornell University Press for advice; our reviewers at the proposal stage; and the final reviewer, Erica Dunn, whose comments and editing were detailed, critical, and insightful, helping immeasurably to shape the final product. Several people helped in more specialized ways during the process of developing the manuscript. Tom Fredericks, Jennifer Shirk, and Ben Zuckerberg deserve special mention in this regard. Much of our thinking has been a consequence of activities and projects undertaken with funding from the NSF Informal Science Education Program, especially the conference on program development and evaluation that we held at the Cornell Lab of Ornithology in 2007 (NSF DRL [ISE] #0610363). This brought in many creative people, each of whom helped us refine a program model of citizen science, with some going on to make contributions to the Center for Advancement of Informal Science Education's citizen science working group white paper, which we cite frequently in this book. Finally, we thank our chapter authors for their enthusiasm and dedication to completing this project.

With special appreciation of The**Cornell**Lab of Ornithology

CITIZEN SCIENCE

Introduction

Why Citizen Science?

JANIS L. DICKINSON AND RICK BONNEY

In this book we examine citizen science within the modern context of the Internet's impact on environmental science, focusing on the burst of large citizen science projects that now engage people in tracking biological and environmental change over broad geographic regions. Such activities are vital to discovering and projecting the impacts of environmental pollution, land use change, and global climate change on species extinctions, distributions, compositions, and ecosystem health. They also provide new opportunities to motivate the public and professionals interested in science-based conservation to work together to expand the shared knowledge base and explore solutions.

Citizen science is a term that, as far as we know, has yet to appear in any official dictionary. But an Internet search for the phrase yields thousands of projects ranging from breeding bird atlases to aquatic insect counts, from frog-watching projects to reef fish surveys. By our definition of citizen science as public participation in organized research efforts, hundreds of thousands of individuals around the world are "citizen scientists," people who have chosen to use their free time to engage in the scientific process.

Here we show how citizen science can be used to study the natural world on broad geographic scales, and describe the critical scientific perspectives and tools required to conduct large-scale citizen science research. We also explore the areas of human endeavor and research that have been integrated with and continue to be influenced by citizen science, the most salient of which is education. Our aim is to provide examples and insights that illustrate the relationship between goals, program designs, and outcomes, including new ideas for program developers, researchers, and educators who wish to work with large-scale citizen science. As citizen science is cropping up in nongovernmental organizations, universities, community organizations, and classrooms, this is a good time to take stock of what we have learned and where the field is going.

Although the large geographic-scale projects described in this book ask participants to contribute effort and data toward common research goals that have been chosen by a centralized team of scientific researchers and educators, many of the ideas translate to program development at smaller geographic scales. For example, we feel strongly that citizen science projects should engage professional researchers at the outset to ensure that there is an authentic intention to publish results based on project data. The balance between research, recruitment of new audiences, and educational goals varies among projects, but if researchers have no interest in the project, or in analyzing its data, then we would consider a project unsuccessful.

Just where projects fit on the spectrum of "real," publishable research varies. Some projects collect snapshot data, good for detecting patterns that are the citizen science equivalent of descriptive natural history observations. These observations have a place in the scientific method if scientists are interested in using them to develop hypotheses that can be tested with more rigorous approaches, so we consider such efforts to be valid citizen science projects. Many projects that seek to recruit new audiences fit this mold. Other projects allow a range of protocols, aiming to move people from less to more rigorous projects and methods; in such cases perhaps only a subset of the data are analyzable with traditional statistics, but the process serves to grow the project and its participant base to achieve longer-term research gains as participants increase their skills and begin to contribute more reliable data. Still other projects pose simple questions for new audiences but use stringent protocols, such as the repeated measures design used by Celebrate Urban Birds, and educate intensively to avoid bias. The essential ingredients are authentic engagement of researchers in the research design and a genuine intention to use the project data to address questions of value to the scientific community with results that stand up to peer review.

Volunteer-based research did not originate with the complex of goals that we write about today. Many projects produce significant impacts on geographical ecology without considering educational impacts on the project participants. Of course, education of participants can improve data quality, especially for projects delivered over the Internet. For example, tests of knowledge and ability, when tied to data on birds, allow researchers to incorporate individual measures of observer skill (or observer bias) in their analyses (Dickinson et al. 2010). But the shift from scientists developing basic monitoring projects to embracing education and community development happened gradually, with increased interest in the interaction between social and ecological systems. As the goals of environmental education are broad, program developers at the Cornell Lab of Ornithology (the Cornell Lab) saw the potential for citizen science to promote the mission of conservation by including, not just skills and knowledge, but attitudes and behavior (e.g., stewardship) as desired outcomes. This increasingly complex view of what citizen science can be informs the structure of this book.

Successfully achieving a wide spectrum of impacts for both science and education requires attention to design principles, resources, and technology. Models developed at the Cornell Lab of Ornithology and elsewhere address a range of issues with regard to project design (Dunn et al. 2006; Bonney, Cooper, et al. 2009). In Part 1 of this book, the authors examine the design of projects aimed at studying the natural world, using examples with a range of goals and audiences. The contributors to Part 2 describe the critical scientific thinking and tools required to conduct large-scale geographic research with citizen science. In the book's final section, Part 3, the authors probe broader areas of human endeavor and research that have influenced and stand to be influenced by citizen science. This section includes chapters about citizen science as a collective action supported by social networks (Chapter 15) and as structural support for community resilience in the face of environmental disasters (Chapter 16).

In a sense, we present large-scale citizen science as a microcosm of the changes that are occurring in science and society owing to the explosion of the Internet and the social web. We ask how citizen science harnesses these changes and what new areas of human understanding might benefit from citizen science in the future. While the principles we discuss are widely applicable to a diversity of scientific topics (e.g., astrophysics with Galaxy Zoo) and social impacts (e.g., human health), we focus on environmental science because it offers a rich history of public participation around the world and because it is such an important arena for public engagement from the standpoint of science and society.

This book highlights some of the learning that has taken place during the past two-and-one-half decades (1987–2012) of large-scale citizen science expansion, and reviews new techniques and ideas that bridge the gap between the social and scientific dimensions of citizen science. We explore the methods, challenges, and underlying goals of citizen science in collaboration with a diverse group of ecologists, conservation biologists, educators, and social science professionals, to provide a novel understanding of what citizen science has accomplished, its projected impacts, and future possibilities. Of particular interest are impacts on environmental behavior, health, community resilience, conservation science, quantitative ecology, and public understanding of science, particularly as they relate to biodiversity and the environment. The central question of this book is: Given that citizen science is a natural fit to combined social and ecological systems, how might it best meet its potential to achieve ambitious individual, societal, scientific, educational, and management goals over broad geographic and temporal scales?

A Brief History of Citizen Science

In North America, the earliest published information about ecology and natural history came primarily from contributions of "amateur" naturalists, many of whose names are quite familiar—Henry David Thoreau, John Muir, John Burroughs. These pioneering scientists had to be self-directed because their lives predated most of the formal science programs that blossomed at colleges and universities in the twentieth century. Today, while many amateurs still dedicate their lives to the individual pursuit of scientific discovery or invention (Howe 2008; www.instructables.com), citizen science has begun to harness the Internet to cover large geographic areas and merge the observations of thousands or even hundreds of thousands of natural history hobbyists into databases that can provide answers to important questions about the range-wide abundance, distribution, movements, annual cycles, behaviors, and natural history of various plants and animals. This Internet-based citizen science is fueled by knowledge that urgent, large-scale questions about environmental patterns and change can be answered only by combining the observations of numerous observers across large geographic areas.

In the past decade we have seen a dramatic increase in the development of new citizen science projects using the Internet (www.citizenscience.org), and many of these focus on animals and plants in the context of conservation. But public engagement in science is not new, and even large-scale engagement has a long history. Volunteer bird surveys began in Europe in the eighteenth century, and North American lighthouse keepers began collecting data about bird strikes in 1880. A group of amateur astronomers started the Astronomical Society of the Pacific in 1889; the National Weather Service Cooperative Observer Program began in 1890; and in 1900 the National Audubon Society began its annual Christmas Bird Count, which more than a century later still takes place at hundreds of locations throughout the United States and Canada each year. Indeed, throughout the twentieth century, individuals across North America participated in projects to count birds (see Figure i.1), reptiles, amphibians, and fish; to monitor water quality; and to scour the night skies for new stars and even galaxies. Thus the recent burst of citizen science project development in North America is building on a long tradition of organized data-collection projects reaching back more than 100 years. We speculate that the growing interest in citizen science is fueled by the merging of ecology and information technology, which has increased the capacity for efficient data collection and visualization, providing a deeper, more immediate form of engagement with large-scale environmental research than ever before.

As interest in citizen science has increased, researchers have begun to define various models for public participation in scientific research (PPSR). Most are based on the type of scientific outcome desired (Cooper et al. 2008)

Figure i.1. A citizen scientist focuses on identifying and counting birds in Sapsucker Woods, Ithaca, New York.

or the degree of "control" that participants have in the project (Wilderman et al. 2004). Recently, a group of researchers working under the auspices of CAISE (Center for Advancement of Informal Science Education) identified three models for PPSR that focus on the degree to which participants are included in various elements of the scientific process (Bonney, Ballard, et al. 2009). Most projects that are considered to be citizen science by their creators fall under what CAISE researchers call the contributory model, for which participants primarily collect and submit data under the gentle supervision of a sponsoring organization. This model contrasts with the "collaborative" and "co-created" models, in which participants are more deeply involved with analyzing data or even helping to develop the project protocols. In this book the focus is explicitly on contributory citizen science projects, in which data collected by participants are of a scale that would be impossible to achieve with ordinary research teams, computers, or sensors.

Many new contributory projects either strive to or serendipitously involve participants more deeply in the scientific process (i.e., Monarch Larvae Monitoring Project; Zooniverse, Galaxy Zoo). A stunning example of this serendipity is that of Dutch schoolteacher Hanny Van Arkel, who opened up a new area of inquiry in astronomy by discovering a green shadow, or "cosmic ghost," in the image of a galaxy, a phenomenon now known as "Hanny's Voorwerp" (Józsa et al. 2009). What is striking about this example is that it mirrors what happens in a university lab: Ms. Van

Arkel used her powers of observation, prior knowledge, and experience to detect something new and interesting, became curious about it, began to ask questions, and then became immersed in the new questions that arose out of her discovery. In the best of cases, citizen science blurs the distinction between scientists and nonprofessional participants, while at the same time maintaining rigorous scientific approaches to understanding processes and solving problems that neither group can solve on its own.

In the field of environmental science, citizen science projects vary along four major axes: (1) initiator of the project, professional scientists or the public; (2) scale and duration of the project, whether local or global and short term or long term; (3) types of questions being asked, ranging from pattern detection to experimental hypothesis testing; and (4) goals, which include research, education, and behavioral change (e.g., environmental stewardship)

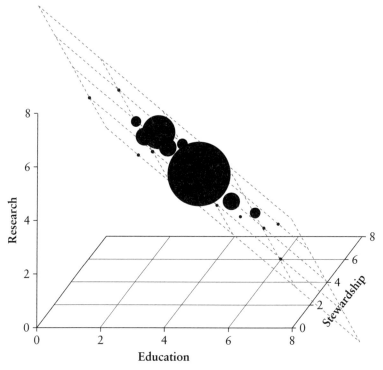

Figure i.2. Trade-offs among the goals of citizen science programs. This figure shows responses to the following question by eighty citizen science project developers who are members of Citizen Science Central's e-mail list when asked to assign weight to the goals of scientific research, education, and promotion of environmental stewardship: "Imagine that you have a pie with 12 slices, and that you have to serve the whole thing. How many slices would you give each of the following project goals (assuming that the biggest goal gets the most pie)?" The dashed grid in this 3-D chart represents a plane through the bubbles and indicates a trade-off between the goals of education and scientific research: as the weighting of education increases, the weighting of research declines.

(Figure i.2). In contrast with collaborative or co-created projects, which often begin with concerns about local environmental problems, projects featured in this book are concerned with problems occurring at regional, continental, and even global scales. Such projects position us to address questions that are not necessarily having visible local impacts. While scientist-driven projects have been described as "using citizens to do science" (Lakshminarayanan 2007), they create "public goods" in the form of important data that can help interpret and conserve the earth's biodiversity (Chapter 15).

What motivates people to contribute to continent-wide citizen science projects? Such projects appear to tap into two basic human tendencies: (1) altruism and interest in collective action toward an overarching goal, and (2) a willingness to collect data of personal value (Van Vugt et al. 2000). Contributory citizen science has thus followed a model of selfish altruism in which the majority of participants likely experience altruistic motivation, at least to some degree, while also receiving tools and resources that support their interests and hobbies.

In some cases, projects begin by exploiting the hobbies and interests of the participants, as with eBird, which is modeled on the desire of birders to keep life lists (Sullivan et al. 2009). If the project is then refined with feedback from the participants, in this case, a large, dedicated community of eBirders (ebird.org), it moves from engineering participants' behaviors and interests (Petersen et al. 2007) to co-engaging project leaders and participants in collaborative design. The consequence of attaining a large buy-in from birding hobbyists is that the data are both voluminous and challenging to work with, necessitating exploration of novel statistical and computational approaches (Marris 2010) and pushing the fields of environmental science and information science into new terrain (see Chapter 8). A secondary benefit is cultivation of an army of talented enthusiasts whose energy can be diverted into increasingly rigorous and refined studies to dissect the ecological processes underlying large-scale patterns (see Chapter 9).

This book acknowledges that projects vary in the emphasis they place on research, education, and stewardship. Emphasis on stewardship stems from a belief that supporting online communities of concern for the natural world will ultimately support both personal action and better policy and management. Practically speaking, one of the key challenges of citizen science is that projects require compromise among the goals of research, education, and stewardship. A survey of eighty project developers (summarized in Figure i.2) demonstrates both variation in their weighting of these different goals and an apparent trade-off or negative relationship between the goals of research and education.

This relationship is worthy of study, but it also speaks to our decision to aim this book at a broad readership. Our intended readership includes practitioners seeking to understand the nuances of how to design or deliver large, Internet-based citizen science projects (Chapters 1–5); how to create research programs using new and existing citizen science data; and how to achieve and study broader impacts of citizen science in the areas of conservation, community resilience, education, ecology, psychology, sociology, and human-computer interactions. Communities engaging with large, public goods projects must wonder: What are the potential impacts for participants, and to what extent can citizen science be integrated with educational and community programming? These are some of the issues we have sought to address and questions we have tried to answer.

Why Cornell and Why Birds?

The long history of public participation in bird studies provides insight into the potential scientific contributions of citizen science, for nowhere has citizen science been more prominent than with birds (e.g., Marris 2010; see Dickinson et al. 2010 for a table of major bird monitoring projects and their impacts). Until recently, bird citizen science projects used paper forms to collect data. In North America this has resulted in hundreds of thousands of historic paper records critical to understanding new problems, such as global climate change (Robbins et al. 2006). Success with projects such as the Breeding Bird Survey, the Christmas Bird Count, Cornell's Nest Record Card scheme, and Project FeederWatch led to a desire for faster and easier ways to get data computerized and available for analysis.

FeederWatch was the first citizen science program to use computer-scannable data forms (1987), but the defining moment leading to the burst of interest in citizen science we see today occurred when it was suddenly possible for participants to transfer data over the Internet. This enabled some of the earliest models of *crowdsourcing*, a term coined to describe the Internet's capacity to enlist the public's help in performing tasks or solving problems that require dispersed effort by a large number of people (Howe 2008). Online data entry was quickly adopted by the National Audubon Society's Christmas Bird Count and the Cornell Lab's Project FeederWatch in 1997, and in the following year by their shared project, the Great Backyard Bird Count. At about the same time, the Breeding Bird Survey of the U.S. Geological Survey (USGS) began to place static, interpolated maps on its Web pages so that researchers and participants could visualize geographic trends.

The Cornell Lab then deployed the very first implementation of a Web-based map as a mechanism for designating locations and collecting geographically referenced data over the Internet. Map-based data submission and storage in a relational database suddenly allowed the development of Web applications that would allow participants to slice and dice the data by location, species, and time period to obtain lists, graphs, and maps to match their interests and curiosity, literally just minutes after submitting the data on which those lists, graphs, and maps were based. This two-way connection, which closed the loop between the remote data collector and the lab, paved the way for creation of research and learning environments that are rich, centralized, scientific, and social. These scientific learning environments have begun to reach out and engage people in studying conservation issues across the planet. We currently estimate that 200,000 people participate in our suite of bird monitoring projects each year. The large volumes of historic and recent data gathered with unprecedented spatial coverage are why bird citizen science data have been analyzed so extensively and why they are deemed so important.

Although in the last ten years citizen science projects have sprung up around the world, the progress of citizen science and the models for creating and delivering successful projects at this scale still come largely from program developers, educators, information scientists, and ecologists focused on birds. Similarly, reviewing the research outcomes of bird citizen science projects is broadly illuminating with regard to the challenges of asking important and relevant questions with large, spatial data sets that, to be used properly, require sophisticated understanding, not only of statistics, but of the data themselves (Dickinson et al. 2010; Marris 2010).

Impacts on Science and Society

While citizen science has produced a plethora of projects and tools to engage people over the Internet, it also has knit together ideas and practices that span the natural, information, and social sciences fields. As examples of informal science education or "free choice learning" (Falk and Dierking 2002; National Research Council 2009), many large citizen science projects carry knowledge from all of these disciplines into projects that engage people to contribute time, energy, and resources to learn about, monitor, and witness the consequences of environmental change. How and why people become engaged in Internet-based citizen science and whether they benefit from participation have become research topics in themselves.

We recognize that citizen science is a moving target. The open source environment is currently developing at a rate that outpaces our understanding of its potential impacts with online projects such as "Make" and "Instructables" that allow people to showcase their own inventions. Computer scientists interested in human-computer interactions have begun to use games and contests to engage the public in discovering new structure for proteins (fold.it/portal) and teaching the Web itself to become more efficient (www. GWAP.com). More profoundly, the social web generates new possibilities for grassroots, collaborative projects (Powell and Colin 2008, 2009), further narrowing the gap between top-down and bottom-up project designs. In the small world of the Internet's social networks (Kleinberg 2000), it is likely that the distinction between local and global will blur as small, localized projects merge into larger, networked endeavors that make use of the principle of collective intelligence (e.g., The Polymath Blog, polymath projects.org).

Scientific empowerment portends large shifts in the form and nature of scientific endeavor and the nature of scientific authority (Irwin 1995). We argue that this is all the more reason to invest time and resources in creating and supporting powerful experiences that promote understanding of science, technology, engineering, and math, hoping that habits of scientific inquiry will develop to promote scientific decision making from a position of knowledge and skill rather than of powerlessness. The rapidly changing landscape of open source Internet applications and tools makes this an exciting time to consider what has been done, what could be done better, and what citizen science might look like in the future.

Direct involvement of professional scientists in citizen science necessitates an expansion of the culture of academic science, which is concerned mainly with academic success (teaching and publication) and citizenship within the professional scientific community. Because it embraces the public's capacity to contribute to the overall equation of what we know about the world, citizen science may have impacts on both scientists and the public. While participants donate their leisure time to learn protocols and biological information or to collect and explore data, professional scientists must navigate new challenges of working to train novice participants; tackling large, complex databases; and learning new statistical techniques, potentially sacrificing scholarly productivity in favor of broader goals related to understanding what needs to be conserved. In this sense, citizen science has the potential to build important bridges between professional scientists and the public with positive outcomes for both science and public scientific literacy.

As Web applications and tools become increasingly flexible, people take matters into their own hands to address key problems of interest. Online groups have already begun to revolutionize public participation in science around specific topics of interest. For example, patients have organized online efforts to compare and track symptoms for diseases that were otherwise unknown and neglected by medical research (Arnquist 2009); in 2008, a Google Earth application populated with modis data and comments from citizen fire spotters provided more up-to-date and comprehensive information than official sources in California's Basin Complex Fire; and information from local water monitors has challenged official reports, influencing policy (Danielsen et al. 2005). We have only begun to see the consequences of this "self-assembly" of scientific knowledge, information, and investigation.

Citizen science draws people into the outdoors to collect data on enchanting organisms, and engages their scientific interest close to home, adhering to "sense of place" (Sobel 2004). Although evidence suggests that participants in some projects may begin to "think scientifically" (Trumbull et al. 2000), citizen science has not been well studied for its potential to change peoples' perceptions of science and of themselves as scientists. Because professional scientists are new players in the citizen science collaboration, it also makes sense to ask how *they* change with involvement in citizen science.

Most well-meaning environmental scientists likely arrive at the doorstep of citizen science with a deficit view of the public's scientific potential, a sense that science is *the* way of knowing, and the notion that they (the scientists) will magnanimously bring scientific wisdom to a hungry or even ignorant public (Irwin 1995). The outcome may be far more interesting than originally perceived. Although scientists who engage with the public may initially buy into the merits of combining education and research for the public good, their involvement in citizen science may cause them to challenge the deficit model and other preconceptions they have about participants. In this regard, citizen science is fertile ground for change in how scientists view the public and, ultimately, how scientists view themselves.

There has never been a time more urgent for developing practices that engage broad audiences in science and technology. With public understanding of science concepts and processes disturbingly low (Durant et al. 1989; Miller 2006), citizen science has a crucial role to play and this role requires investigation. Increasing the public's ability to think critically about science is perhaps the only way to diminish the impact of information put forth by corporate entertainment news sources, which are rife with misrepresentations of important scientific and political issues. These misrepresentations have already begun to influence public attitudes and behavior with respect to important concerns such as public perception of the reality of anthropogenic causes of global climate change (Boykoff 2007; Boykoff and Boykoff 2004).

Citizen science projects designed within the contexts of biodiversity monitoring and environmental change provide a natural bridge to connect audiences fascinated with science and technology with their polar opposites, people who love nature, but fear the effects of modern values and lifestyles on the environment and blame science and technology for the bulk of the environmental problems we face. Scientists' concerns about public denial of the severity of current and future environmental crises are somewhat ironic, for scientific progress is in some sense to blame for global contamination of people and nature. The preponderance of progress-centric ideologies and failure to act indicates that scientists themselves are capable of scientific denial (Beck 1995; Dickinson 2009). At a time when we are facing serious risk due to environmental degradation, habitat loss, and climate change, it is difficult to imagine a more fruitful place for cross-talk than the environmental context, where transformation of public psycho-spiritual values has played a critical role in raising environmental awareness and concern in the first place (Kovan and Dirkx 2003). It is the potential for bringing diverse ideological groups together to tackle urgent problems of common concern that makes citizen science the compelling endeavor that it is today.

A Vision for Citizen Science

What claims are currently made for citizen science? First is the assertion that designing projects with multiple goals will ensure outcomes that extend well beyond scientific impacts. Not only can scientists answer important questions with citizen science, but engaging the public in research is thought to have a number of potential societal benefits. The hope is that citizen science leads to increased scientific literacy (Krasny and Bonney 2005), increased interest in and knowledge about a range of environmental issues, and increased capacity for people to assemble the tools and data needed to move toward a level of scientific understanding that promotes autonomous, informed choice. The hunch is that having an engaged, educated, nature-loving public helping to drive environmental policy is perhaps even more critical to addressing public conservation problems than further research is.

Recent "systems thinking" approaches to ecosystem ecology suggest that citizen science will come to play a vital role in helping to address the complexity of human and natural systems, which vary across space, time, and organizational units (Machlis et al. 1997). To the extent that ecological systems can be understood only through collaborations involving both environmental and social scientists, citizen science is the perfect tool. It provides ready-made opportunities for dovetailing research on human behaviors with research on ecological phenomena, new and historic, to understand feedback loops, time lags, and outcomes that would be impossible to track without combined data on human and natural inputs. These approaches are rare enough, but are basically unknown at the large, geographic scales possible with citizen science.

While continental-scale biological monitoring has been a major focus of citizen science in North America and the United Kingdom since the early 1900s (Greenwood 2007a), we have barely scratched the surface in understanding the impacts of citizen science on participants and on scientists themselves. In the first decade of the 21st century, research methodologies began to move across disciplines with the idea that social scientists can analyze human impacts using landscape-based frameworks and statistical approaches similar to those used by landscape ecologists (Field et al. 2003). We are also seeing a variety of ways in which ecological understanding is enhanced by incorporating information about the human social landscape

(Michaelidou et al. 2002). For example, if socioeconomic status and ethnicity influence landscaping practices, as demonstrated in a study conducted through the Phoenix Long-Term Ecological Research site, then the social landscape may be an important predictor of local bird or insect concentrations, creating a mosaic of patterns of abundance that is inexplicable without information on the spatial diversity of human practices within the ecosystem (Lerman and Warren 2011). This simple example illustrates the value of collecting biological data simultaneously with data on human practices.

Our hope for citizen science is that these possibilities will be augmented by putting into practice new understanding of citizen science goals and project designs, including incorporation of social networking tools, which foster the building of networked communities with learning and conservation impacts that we have only begun to imagine (e.g., www.YardMap.org). The potential impact of citizen science on people is arguably as important as its impact on science, making this an ideal time to investigate the theory and practice of citizen science from an interdisciplinary perspective. The combination of social sciences approaches with citizen science research and project design points to a rapidly changing frontier that will undoubtedly break new ground. What better time to take stock and investigate the new directions in which this field is heading?

In some sense, this book is a call for further exploration of the limits of citizen science. Broadening the scope of impacts beyond research and education forms a natural bridge between citizen science and new studies linking engagement with nature to issues of environmental justice, diversity, and human well-being (Wells and Evans 2003; Wells and Lekies 2006; Wells 2000). Citizen science is poised to support the shifts in education policy and thinking that characterize the children and nature movement. The fit of citizen science to this new movement is obvious, but just how we use citizen science to reach large segments of the public is an interesting problem (Louv 2005). As citizen science is a relatively new endeavor for scientists, educators, and most participants, it is in need of scrutiny, academic exploration, and synthesis if it is to achieve its cross-cutting goals and impacts.

Every project discussed in this book aims to get people outdoors to reconnect with the natural world with potential consequences for our collective future. If, for the first time, our survival depends not on how fast we progress but on how we handle risk (Beck 1991), then resilience will be crucial for this next generation, and to the extent that citizen science and myriad other ways of reinstating the children-nature connection have potential to build resilience, we should invest in them. If Louv (2005) is right, the social imperative is to learn more about how to get people, particularly families and children, out into the natural world, where they will begin to know it and care for it; the environmental imperative is to get people of all ethnic and economic backgrounds on board with what we need to do to save what we can of the natural world; and the scientific imperative is to figure out what it is that can be done. It is difficult to imagine how to move forward without an infrastructure like citizen science for putting these pieces together. It is with this in mind that we have called upon the authors of this book to share their visions of citizen science now and into the future.

PART I

THE PRACTICE OF CITIZEN SCIENCE

Developing, implementing, and evaluating the impact of citizen science projects is a complex endeavor. How can projects be designed to accommodate potentially competing goals and objectives for research, conservation, and education? How can projects be organized at the vast geographic scales necessary for understanding environmental problems that stretch across continents and around the globe? How do we ensure data integrity and quality data analyses? Given our goal of large geographic coverage and repeated observations at the same site, how do we recruit and train large numbers of participants to donate their time to participate in essentially altruistic endeavors? And how can we measure the impacts of citizen science, both for project participants and for society as a whole? The first section of this book tackles the overarching challenge of project design, implementation, and assessment by tapping into the expertise of practitioners who have been building citizen science projects and tools for many years.

Chapter 1 by Bonney and Dickinson is an overview of citizen science project development that focuses on the strategies that have evolved over the years at the Cornell Lab of Ornithology (Bonney, Cooper, et al. 2009). The Cornell Lab's approach to citizen science is to develop a spectrum of projects on birds that balance our goals of conservation research, science education—including reaching out to diverse audiences—and stewardship. The model describes the steps and stages that must be considered when developing a project to produce robust outcomes for research and education.

To help the program model come alive, Chapter 2 describes four ongoing citizen science projects developed at Cornell and beyond. The four case studies explore projects that have been designed with unique scientific goals, educational objectives, and intended audiences, and which are carried out at varying scales and levels of complexity. Two of the projects (FeederWatch and Neighborhood Nestwatch) focus on birds, which are charismatic and set a high bar in terms of project potential, with respect to both the quality of data that can be gathered by talented amateurs and the number of participants who stand ready to be recruited. Along with reef fish and celestial objects, birds may be one of the few targets for which the data-collection skills of amateurs outpace, or at least compare favorably with, those of professionals. This chapter also highlights successful projects from the world of plant phenology (Project BudBurst) and insect ecology (Monarch Larva Monitoring Project).

The changing technological landscape continues to present new possibilities as well as challenges for project developers. In Chapter 3, Kelling explains the practical aspects of creating a sound cyberinfrastructure to serve as a backbone for large-scale citizen science project development. In the process he describes an array of strategies for creating novel and enduring applications to deliver citizen science and its data to project participants, professional scientists, and managers over the World Wide Web. The strategies include recent improvements in Web application design for interactive maps, graphing, and online data analysis tools.

While birds can be considered charismatic, charisma is not necessarily required for success, for we have seen successful new citizen science projects focused on species like grunions (Grunion Greeter, grunion.org) and earthworms (The Great Lakes WormWatch, greatlakeswormwatch.org). And even though birds are charismatic, Cornell's bird projects, with just 200,000 participants, have recruited only a small percentage of the estimated 46 to 70 million birders in North America (Cordell 2004; USFWS 2002). This suggests that the outreach potential, even for birds, may be much greater than current project strategies achieve. We still have much to learn about recruitment of participants.

Successful implementation, maintenance, and growth of citizen science projects cannot be accomplished without expertise on "getting the word out." In Chapter 4, Chu et al. examine insights that have arisen during fifteen years of marketing and communicating about citizen science projects based at Cornell. The critical issue of project sustainability must be considered even during project development, because although the scientific value of projects increases with project duration, sponsors typically fund new projects rather than ongoing ones. What's more, scientific funding clusters are just beginning to see citizen science as a serious and fundable research endeavor. Therefore, sustaining projects requires tapping into a variety of funding opportunities focused on educational, cyberinfrastructure, computational, conservation, social networking, and even environmental psychology research. This chapter provides ideas for practitioners and institutional directors facing the sustainability challenge.

In the final chapter of this section Phillips et al. focus on citizen science program evaluation (Chapter 5). Because many citizen science projects are designed with specific educational objectives, we are concerned about the lag in research required to understand project impacts within the context of these goals. Do individuals feel more empowered to evaluate scientific information or to make science-based choices as a consequence of participating in citizen science? When individuals are exposed to citizen science experiences through an organization, rather than of their own volition, do they emerge from the experience with new ideas about themselves as scientists or about the value of science to society? This chapter reviews recent progress made in the evaluation of informal science education projects and defines expected outcomes, setting the stage for future evaluation and assessment research.

1

Overview of Citizen Science

RICK BONNEY AND JANIS L. DICKINSON

While the term "citizen science" may be relatively new, the idea that any person can participate in scientific research—regardless of background, formal training, or political persuasion—is as old as Aristotle. After all, scientific knowledge is largely derived from observation, experimentation, and analysis, which most people are capable of, at least at a basic level. Basic scientific observations, such as kinds, numbers, and locations of plants and animals, can be made and recorded by anyone who carefully examines the world around them. Furthermore, such observations can lead naturally to new discoveries, insights, and questions. Participants in the Cornell Lab of Ornithology's Project FeederWatch, for instance, were the first to report the emergence of a disease (conjunctivitis) that eventually spread throughout the House Finch population in the eastern United States and caused a significant decline in House Finch populations (see Chapter 2).

Whereas many projects now involve the public in various aspects of scientific inquiry, the projects in this book generally follow the model for large-scale, Internet-based project development and implementation developed at the Cornell Lab over the last two decades and refined at a conference of citizen science practitioners who gathered at the Lab in June 2007 (Bonney, Cooper, et al. 2009). The model differs from earlier approaches (e.g., Dunn et al. 2006) in focusing on geographic extent of data collection and taking a cross-disciplinary approach to program development by assembling teams to work at the interface of education, environmental biology, social science, geospatial statistics, evaluation research, marketing, communications, and information science. Such teams then develop and implement protocols for rigorous data collection and submission that will engage motivated individuals who may or may not be formally trained in natural history observation (e.g., bird-watching) or scientific investigation. The model is highlighted by the NSF-sponsored Citizen Science Toolkit website (www.citizenscience.org), which was developed after the 2007 conference, and this model of citizen science is expanding continually as new ideas, insights, and tools are added by the burgeoning citizen science community.

Environmentally focused citizen science projects ask people to observe and collect data about plants and animals and to submit their data to centralized databases. Projects have been developed to study animal migrations, bird nesting behavior, populations of fish around coral reefs, breeding behavior of grunions on California beaches, and even earthworms around the Great Lakes (see project listing at www.citizenscience.org and table describing the primary bird monitoring projects in Dickinson et al. 2010). Typically projects are developed and managed by universities, museums, or environmental/natural history organizations with a direct interest in using the data for research, such as the Colorado Climate Center at Colorado State University (the Community Collaborative Rain, Hail, and Snow Network), the Boston Museum of Science (Firefly Watch), and the National Audubon Society (Christmas Bird Count). The project team develops and supplies project instructions, data sheets, and generally a website for reporting observations.

Project participants range in age from young students to senior citizens and come from varied backgrounds. What they have in common is a strong interest in the organisms being studied, a curiosity about the world around them, and a desire to advance the field of science. Many participants learn a lot about scientific investigation through their data collection endeavors, and some go on to conduct their own studies and data analyses. According to Karen Oberhauser, director of the Monarch Larvae Monitoring Project, many children who find and count monarch caterpillars experience what she calls "science bonding" and go on to conduct experiments about the relationships between monarchs and their host plants, the milkweeds.

As participants in citizen science projects submit data to project databases, scientists at the sponsoring institutions (or their partners) analyze the accumulating information and report findings on project websites and in popular and scientific publications. Vetting, cleaning, and analyzing the massive databases that result from large projects is a complex job and requires substantial training in statistical analysis and informatics. However, the payoff can be huge as the data reveal environmental patterns and processes that can be understood only through data collected across large geographic areas and over long periods of time. For example, citizen science data are now revealing northern range shifts in a variety of species as an apparent result of global climate change (Root 1988; Zuckerberg, Woods, et al. 2009).

NestWatch

Multiple Nest Visits Data Sheet

This form is for your records — use it to locate and describe your nest site and record up to 10 visits to a nest. Use a separate form for each nest monitored and each new nest attempt. See keys on the next page for explanation of codes and fields. If response is "Other" enter "OT"; for "Unknown" enter "U."

Year_____ Species _____

1. Nest Site Location (see key on next page)

Nest Site Name:
_____

Nest Location (nearest street address OR lat/long):
_____
_____

OR
    Latitude N _____
    Longitude W   – _____
Zip Code _____

2. Description (see key on next page)

Nest is located (circle one) ☐ IN ☐ ON ☐ UNDER
    Substrate_____
Habitat within one meter _____
Habitat within 100 meters _____
Habitat modifier _____
Elevation (specify ft. or m.) _____
Height above ground (ft. or m.) _____
Cavity orientation_____
Cavity opening width (specify in. or cm.)_____

3. BREEDING DATA *For columns C, D, E, and J, enter "X" if eggs or young are present, but exact number is unknown. Enter "U" for unknown.

Visits	A. Month / Day (1–12) /(1–31)	B. Time (am/pm)	C.* # of Eggs	D.* # Live Young	E.* # Dead Young	F. Nest Status	G. Adult Activity	H. Young Status	I. Mgmt. Activity	J. Cowbird Evidence		K. Observer Initials	L. Comments below
Ex	5 / 4	4 pm	4	0	0	CN	AA	—	No	IE	0	0	MS	✓
1	/													
2	/													
3	/													
4	/													
5	/													
6	/													
7	/													
8	/													
9	/													
10	/													

4. NEST ATTEMPT SUMMARY Fill in the HOST information below after the nest attempt is complete.

IMPORTANT DATES

First Egg Date	
Estimated Hatch Date	
Estimated Fledge Date	

TOTALS

No. of visits	Clutch size	No. Unhatched eggs	No. live young	No. fledged

COMMENTS:

Site Name

Rev. 4/15/08   **PLEASE TRANSFER DATA ONLINE AT www.nestwatch.org. THANK YOU!**

Figure 1.1 Example of the data-collection form, which mimics the online data entry form for the Cornell Lab's NestWatch program (a national nest monitoring project).

Despite the apparent simplicity of the citizen science concept, developing, implementing, and assessing the impacts of public participation in research is in truth a complex and exacting process. To be successful in achieving simultaneous objectives in research, conservation, and education, the project team must plan protocols, recruit participants, manage data, disseminate results, and evaluate outcomes in a deliberate manner. In addition, because most large-scale projects are now delivered online, their development requires careful attention to best practices in website development and management. In the next section we present an overview of citizen science project design. For detailed treatment, see the resource list in the Citizen Science Toolkit at www.citizenscience.org.

Overview of Project Design Considerations

Choosing a Scientific Question

Citizen science projects are ideally driven by a research question or monitoring agenda that fits clearly within the sponsor's scientific or conservation mission. Although a project's main goal might be education, participants need to know that they are participating in "real" science research that will lead to analysis and publication in order to reap the educational benefits of their participation. When choosing questions, project developers must consider that many participants will be amateur observers, at least at first. Therefore, monitoring studies for which data collection relies on basic observation skills, such as counting the numbers of birds at a feeder or locating invasive plants, are the easiest to develop into large-scale citizen science efforts. However, citizen science can involve complex designs and even experiments if developed with continual audience testing and feedback. For example, participants in the Cornell Lab's Birds in Forested Landscapes project select survey sites, describe site habitats, and use playbacks of recorded songs to locate and map breeding birds (see Chapter 9).

Forming a Project Team

In most cases it is necessary to have funding in place prior to initiating a large-scale citizen science project. Citizen science projects with many complementary goals and objectives require a team of developers with varied expertise to be successful. As illustrated by the wide-ranging backgrounds of the contributors to this book, many large-scale citizen science efforts involve scientific researchers, formal and informal science educators, computational statisticians, social scientists, and evaluators to help set learning goals and define intended project outcomes. Of course not all institutions and sponsoring organizations employ individuals with all of the required expertise, which is why so many projects represent collaborations and partnerships among complementary organizations or institutions. Organizations that do not have an explicit plan for scientific research are unlikely to be able to fulfill their obligation to participants to actually use the data. Significant groundwork has been laid regarding how a team might develop detailed scientific goals and methodologies to the benefit of the research effort (Dunn et al. 2006; see also Chapter 10 for an excellent example of this).

Developing and Refining Project Materials

The success of citizen science projects hinges largely on the quality, utility, and flexibility of their support materials, including project protocols, data forms, and educational resources. New, Web-based projects may successfully meet these requirements in part by engaging a knowledgeable user community. User support must be clear, intuitive, and tested repeatedly with the target audience to ensure that accurate data will be collected and submitted and that participants will learn from the process. Even so, there is often an experience effect, wherein data become more usable after participants have had a practice year (see Chapter 6).

Project protocols work best if they are easy to perform, explainable in a straightforward manner, and engaging for project participants. Protocols can be tested by observing participants in the field as they collect and submit data (see Chapter 11). If protocols prove to be confusing or overly complicated, they can be clarified, simplified, or otherwise modified until newly recruited participants can follow them with ease.

Quality data forms (Figure 1.1) mirror the project protocol and the online data-entry form to help facilitate complete and accurate data collection and to help prepare data for analysis. Online data forms, now used by most projects, also require participants to enter all essential information. Such forms can also filter anomalous records before they enter the database.

Educational resources can include identification guides, posters, manuals, videos, podcasts, and newsletters that describe the project and its central questions, including the challenges that observers face in making observations or filling out data forms. Such materials can be developed with deep consideration of the target audience's cognitive biases and with scholarly research on how people learn to perform similar tasks (see Chapter 11). In addition, online forums and social networking tools offer many opportunities for project support that are just beginning to emerge (see Chapter 16).

Recruiting and Training Participants

Locating project participants and convincing them to spend their free time collecting and submitting data is one of the most challenging aspects of implementing a citizen science project. Many projects have developed full sets of materials yet failed to recruit enough participants to gather data sufficient to address the project's central questions. At the Cornell Lab, it is common practice to engage with our communications and marketing group during the project design phase to ensure that projects match not only the altruistic tendencies of our target audience but also their motivations and interests (see Chapter 4).

Providing participants with the support they require to understand project materials and gain confidence in their data-collection skills is critical. Volunteers generally want to know that the data they collect are useful, and routine "quizzing" can be one way of assuring that they are. Quizzes can be fun and informative, while also allowing project leaders to recognize and reward skillful contributions. If testing is necessary or desirable, it can be done most productively by offering opportunities for project participants to learn more and to observe their own improvement.

Accepting, Editing, and Displaying Data

All of the information collected by a citizen science project must be accepted, edited, and made available for analysis by professional scientists as well as the public. Several factors must be considered. A project must find an appropriate data-management platform or design a new one, which can be expensive. Database standards and metadata structure must be determined. Issues of data security and privacy must be addressed, such as encrypting and managing secure data (for example, for locations of endangered species). Data must be unified to be made available to all users and must be protected through secure backups. For all these reasons, data management and manipulation are complex processes and are described in detail in Chapter 3.

Facilitating data exploration by project participants and other members of the public is also critical. Indeed, allowing and encouraging participants to manipulate and study project data is one of the most educational aspects of citizen science. Current Cornell Lab projects allow participants to view a diverse set of graphs, maps, histograms, and other visualizations (for examples, see Chapter 8). Many of the projects also supply personal data-management tools such as those that create sophisticated birding life lists or compare information on breeding success in nest boxes from one year to the next. These tools are extremely popular with project participants and have helped increase the quantity of data submitted (see Chapter 3).

Analyzing and Interpreting Data

Having tens, hundreds, or even thousands of volunteers collecting data can increase the statistical power of research but may also result in large data sets of varying completeness across time and space that require significant expertise to analyze. Dealing with such data can raise both technical and philosophical questions. Fortunately, the large size of many citizen science projects creates a favorable "signal to noise" ratio and yields strong patterns that are reasonably easy to interpret. Even so, organizations that become seriously involved in citizen science should plan on employing staff members who are trained in sophisticated data-management and analysis techniques.

Because of difficulties inherent in estimating and controlling for detectability of the organisms being studied, citizen science data are generally best suited for computing indices of relative abundance rather than for making estimates of absolute abundance. They are also excellent for showing overall patterns of distribution such as range centers and range shifts. Analyzing and interpreting citizen science data is such a complex issue that it is treated extensively in three chapters in this book (see Chapters 6–8).

Disseminating Results

Researchers should place a high priority on rapid analysis of citizen science data, once enough data are on hand. Many questions that interest researchers require significant time series—spanning several years—and this should be communicated to participants. It is also incumbent on researchers to summarize the results of peer-reviewed research for the participants. Finally, it is important to reach out to other target audiences, such as landowners, conservation partners, and managers.

Letting the world know about the results and conclusions of citizen science projects is not only a moral obligation to participants, it is important in recruiting new participants, retaining existing ones, and garnering financial support for project continuation. Results can be disseminated on project websites, in newsletters, in newspaper and magazine articles, and of critical importance, in the scientific literature. The website www.citizenscience.org includes dozens of references to articles published in peer-reviewed literature on both scientific and educational impacts of citizen science efforts.

Also very important is preventing dissemination from becoming unidirectional ("top-down"). Scientists and project staff can receive helpful feedback from project participants regarding analyses and results, and many participants are quite capable of disseminating their own results and generating their own hypotheses. Supporting forums for sharing observations among participants is likely to increase project support and individual learning.

Measuring Impacts

A final step in citizen science project development involves measuring project outputs and outcomes to ensure that both scientific and educational outcomes have been met. If they have, publications and websites can elaborate these successes for others to use as models. If they have not, evaluation can show how to improve a project or design better projects in the future.

Outputs and outcomes can be gauged in many ways. Some measures reflect greater knowledge in scientific fields, some reflect improved scientific literacy among the public, and some reflect both. Measures of scientific contribution can include such items as numbers of papers published in peer-reviewed journals, size and quality of citizen science databases, and frequency of media exposure of results.

Measuring improvement in public scientific literacy is more challenging. Possible measures include participant retention, enhanced participant understanding of the process of science, improved participant skills for conducting scientific explorations, and increased participant interest in science as a career. Another science literacy benefit worth examining is increased support for conservation efforts. Examining methods for measuring the educational impacts of citizen science participation is discussed in detail in Chapter 5.

☙❧

Citizen science has raised more questions than it has answered because it is a constant reminder of the need to make science relevant to the public and of the permeable boundary between amateurs and professionals— that is, between people who are curious about the world and achieve the ability to think scientifically and professionals who are trained to do this for a living. As an endeavor, citizen science is hungry for new tools that touch on a variety of scientific and social science fields, including database management, scientific analysis, learning theory, and educational research. We are already seeing a requirement for innovative and rigorous statistical analysis methods to handle the massive amounts of monitoring data that are being collected. But citizen science has also begun to push the frontiers of understanding with regard to how people learn and how they begin to think scientifically across geographic regions and cultures. Today we see citizen science as open terrain for researchers from a variety of disciplines, whether they are interested in pushing the frontiers of science, engineering, and computer science, or research in the areas of education, psychology, and sociology. This means that citizen science, as an emerging field, is constantly searching for innovation: new language to talk across the disciplines, new models for understanding project design and project results, and new tools to bring people, nature, and computers together in meaningful ways.

Acknowledgments
We thank our colleagues at the Cornell Lab of Ornithology, participants in the 2007 Citizen Science Toolkit Conference, and members of the CAISE Inquiry Group for sharing their insights. This work was supported by the National Science Foundation grant ESI-0610363.

2

Projects and Possibilities

Lessons from Citizen Science Projects

Even with a tool kit handy, understanding how to create, run, and achieve the desired outcomes of a citizen science project can be challenging without detailed information about what types of efforts work well. Specifically, it is useful to learn about the issues that project leaders find most challenging and how a variety of practitioners have gone about the process of creating and delivering citizen science projects to a diversity of audiences and within a wide range of subject areas and contexts.

In this chapter we have asked experienced project leaders to share their insights regarding design and delivery of citizen science projects. The thoughts they offer here represent the cumulative effort and wisdom of a large number of citizen science professionals and other collaborators who together have designed and delivered a wide range of environmental, ecological, and conservation-based projects.

From Backyard Observations to Continent-Wide Trends: Lessons from the First Twenty-Two Years of Project FeederWatch
DAVID N. BONTER

The origins of Project FeederWatch (PFW) date to 1976 in Ontario, Canada, when ornithologist Dr. Erica Dunn of Long Point Bird Observatory (LPBO) designed the Ontario Feeder Bird Survey to monitor changes in the abundance and distribution of birds that visit bird-feeding stations. By 1987 the survey had grown to include more than 500 participants, and LPBO asked the Cornell Lab of Ornithology to partner with them in creating an expanded project called FeederWatch to monitor changes in winter bird populations throughout all of Canada as well as the United States. Over the next two decades the project grew to include approximately 15,000 participants each year with more than 110,000 total checklists submitted annually from all U.S. states and Canadian provinces. More than 45,000 individuals have enrolled in the project since 1987 (Figure 2.1).

The research goals for Project FeederWatch are (1) to document geographic variation and both short- and long-term changes in the winter abundance and distributions of North American bird species that visit feeders; (2) to study factors that cause variation in numbers and distributions of birds, such as foods offered, habitat and landscape features, and weather; (3) to evaluate the use of feeder counts as a method of monitoring bird population trends; and (4) to undertake other research projects that take advantage of the interests and abilities of FeederWatch participants (Sutcliffe and Bradstreet 1994).

As the project evolved, several educational goals were established including (1) to enhance the ability of participants to identify species that visit feeders; (2) to increase participant knowledge and appreciation of local biodiversity; (3) to improve participant knowledge of the steps required to create a safe bird-feeding environment; and (4) to increase understanding of winter bird movements and population changes. Ensuring that research and education goals are all being met has required significant effort to develop educational and training materials that help participants comprehend and adhere to project protocols and increase their overall knowledge of bird biology and ecology.

Project Design

The FeederWatch protocol has remained unchanged for more than two decades, providing an increasingly long-term set of standardized observations. In brief, participants establish a count site centered on a bird-feeding station. Most sites are in residential yards, but some participants monitor feeders at schools, nature centers, senior centers, and other locations. Participants select a two-day observation period as often as once per week from early November to early April. During each period participants note the maximum number of individuals of each species seen at one time within their count site and report that number to the FeederWatch database. This protocol was developed with considerable deliberation, because it is important for observers to avoid repeatedly counting the same individuals (pseudoreplication). Participants also report details about the number of feeders and foods provided, observer effort (measured in cumulative hours and in the number of half days of observation during each two-day count period), and weather conditions (temperature extremes, precipitation, depth of snow cover).

Figure 2.1. Willard and Lucille Smith are two of the 119 original FeederWatchers who joined the project during the first season and participated in each of the first 20 years.

Participant Interaction

On the premise that communication between project staff and participants is essential to fostering continued participation and learning, FeederWatch periodically sends project updates, news, and reminders to volunteers by e-mail. An annual summary of FeederWatch results, *Winter Bird Highlights*, is mailed to all participants in October. U.S. participants also receive *BirdScope*, the newsletter of the Cornell Lab of Ornithology, while Canadian participants receive *BirdWatch Canada*, the newsletter of Bird Studies Canada. Project staff answer participant phone and e-mail questions relating to bird identification, bird feeding, project protocols, online data entry, and confirmation of unusual or unexpected observations. The resources required for project administration should not be underestimated; current staffing includes a full-time project leader and fifty hours of project assistant support each week throughout the year. In 2008 FeederWatch staff responded to several hundred phone calls and more than 6000 e-mail messages. The ease of electronic communication has led to an increase in interactions between staff and participants, with the volume of e-mail messages received doubling between 2004 and 2008.

Interactions among participants began in 1999 with the introduction of PFW-L, an e-mail discussion group designed to provide a venue for participants to ask each other questions and share knowledge concerning birds and bird feeding. This group allowed participants to build a sense of community around the FeederWatch experience. During the 2008–2009 season, a new Web-based networking system called the FeederWatch Forum was initiated to allow users to expand their social interactions and, for the first time, exchange photos. More than 5000 posts were contributed by 225 users in the first year. Dozens of forum threads cover topics such as tips for attracting birds and seasonal changes in bird observations, and offer assistance with making challenging identifications. The forum is moderated and maintained by the community and requires minimal input from project staff.

Training and Educational Resources

FeederWatch staff have developed numerous educational resources to help project participants learn about birds and successfully contribute accurate data. These include printed materials that are mailed to all FeederWatch participants at the beginning of the season (the "research kit") and online resources available to anyone whether or not they participate in the project.

The research kit includes several items, starting with a letter containing the participant's identification number and important dates and reminders. An instruction booklet provides clear, simple instructions for recording and submitting FeederWatch observations via computer-readable paper forms or through the online data-entry system. A full-color identification poster features images of species likely to be seen in the participant's region, providing a basic resource to assist with bird identification. *The FeederWatcher's Handbook* describes the types of feeders and foods that are typically offered to wild birds and provides advice on how to design a backyard feeding station and deal with typical problems such as squirrel damage and disease outbreaks. Finally, the research kit includes a calendar featuring photos and quotes submitted by participants during the previous FeederWatch season. The entire kit is mailed to all new FeederWatchers, while returning participants receive a subset of the materials (or can choose to save paper and obtain what they need online).

Web-based training and educational resources, which include all of the materials found in the research kit and much more, are constantly being revised and expanded in response to participant feedback. Among the most popular pages are those on "tricky bird identifications," where photos and detailed comparisons provide users with the information necessary to distinguish among similar and confusing species.

Multimedia elements on the website are increasingly valuable tools for conveying information to project participants. For example, Lab staff filmed an instructional video at the home of a project participant in 2008. The video walks prospective participants through the various steps of the FeederWatch protocol, providing a set of visual instructions to supplement the online and printed materials. The video streams through the Feeder-Watch website and was viewed 6,700 times during the first year.

In addition to instructional and educational resources, numerous on-line tools have been developed to allow any visitor using the website to answer questions about the distribution and abundance of birds using the FeederWatch data set. FeederWatch Web pages were accessed nearly 1.1 million times between June 2010 and June 2011. The FeederWatch web-site contains more than 1,500 pages; approximately 60 pages include sup-port materials and more than 1,200 pages provide opportunities to explore the FeederWatch database. The "Explore Data" section includes tables of the most common birds reported by state, province, or region; dynamic animated maps showing changes in the numbers and locations of birds over time; population trend graphs; and photos from confirmed rare bird reports. In addition, registered participants can access their own personal data summaries and compare their observations among seasons.

Data Collection and Validation

Receiving FeederWatch data and ensuring that the information submitted by participants is of high quality remains an ongoing challenge. When the project began in 1987, all data were submitted on computer-readable paper data forms, a first for a citizen science project, and an innovation that made it possible to get results out before the next winter season began. The forms were scanned, and archived in an Oracle relational database. In 1999, on-line tools were developed that allowed participants to enter their data di-rectly into the Oracle database via the Internet. Over the past decade the online data-submission system has undergone a series of upgrades, and in 2011 more than 80% of participants submitted their data over the Web.

Online data entry is not only convenient for the participant; it also al-lows data to be automatically validated in real time through a set of geo-graphically based filters that flag unusual reports. These flagged reports are potential errors that require the user to "confirm" their observations before proceeding. When a participant does confirm an unusual report, the observation is forwarded to an expert for further review. For example, when a Rustic Bunting, an Asian species rarely seen in North America, was reported by FeederWatchers Harvey and Brenda Schmidt in Creighton, Sas-katchewan, in 2009, the flagging system automatically flagged this highly unusual report for review. The Schmidts were asked to confirm their re-port, and they submitted several definitive photos that confirmed the first-ever record of the species in the province. As this example demonstrates, the validation system often allows participants to provide supporting documentation when the bird in question is still present at the observation site. Flagged reports that lack verification are excluded from data analyses and Web-based data tables, but remain on the participant's personal checklists. The system also allows researchers to identify those volunteers who are in need of support and to provide additional assistance where it is needed, ultimately improving data quality and integrity.

Despite advances in automated data validation, the online review system is not perfect. Although it flags unexpectedly high counts of birds and highlights reports of birds that are outside their typical range, it fails to flag incorrect but plausible reports (e.g., misidentification of a bird as another species that would be expected to be found in the region). Screening data for these errors remains a challenge. Results for easily misidentified species can be presented with caution. For instance, data on the easily confused Black-capped Chickadee and Carolina Chickadee are lumped for analyses in the zone of overlap among these species. The ultimate solution to identification challenges, however, likely lies in additional support and education of participants in order to prevent such errors in the first place.

Impacts

With more than 1.6 million checklists submitted during the first 22 years of the project, FeederWatch data have been used to explore several areas of scientific inquiry. The case of the Evening Grosbeak, a widespread finch, provides a dramatic example of how FeederWatch data have been used to track changes in the abundance and distribution of winter bird populations. When the project began in 1987, Evening Grosbeaks were among the most common birds reported at feeders across much of North America. Within just a few years, however, participants began to report a decline in the numbers of grosbeaks. One participant in Connecticut wrote, "I really miss the Evening Grosbeaks. When we started FeederWatching, we would get more than 30 on the feeders, but we have not seen one in more than four years." A FeederWatcher in Vermont noted, "We have noticed a few small changes over the years, but one that has always puzzled us is that when we started 20 years ago, Evening Grosbeaks were consuming 50 pounds of sunflower seeds per week, and now it has been several years since we have seen even one." Analysis of FeederWatch data revealed that the proportion of sites reporting Evening Grosbeaks plummeted by 50% in eighteen years. At locations where the species continued to be seen, mean flock size declined by 27% (Bonter and Harvey 2008; see Figure 2.2).

Additional FeederWatch publications have focused on sources of bird mortality in the backyard such as predation by hawks and cats and the effects of window collisions (Dunn 1993; Dunn and Tessaglia 1994). In the 1990s, a series of publications documented previously unexplored winter movements of various species at the continental scale (Hochachka et al.

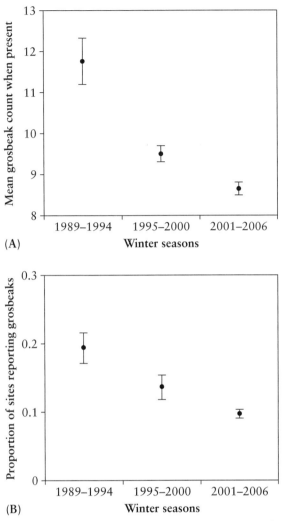

Figure 2.2. FeederWatch data showing long-term declines in Evening Grosbeak populations. (A) Mean Evening Grosbeak flock size when present at a site (B) Average proportion of Project FeederWatch sites in the United States and Canada reporting Evening Grosbeaks at least once during a winter season. Values for individual winters were averaged across six seasons (mean and 95% CI). Originally published in Bonter and Harvey 2008.

1999; Wells et al. 1996; Wells et al. 1998). The influence of West Nile virus on bird populations has also been investigated, revealing a profound effect of the disease on crows and their relatives following the initial spread of the virus in North America (Bonter and Hochachka 2003). In 2010, FeederWatch data were linked with land cover information to examine the colonization of North America by an invasive, nonnative bird, the Eurasian Collared-Dove (Bonter et al. 2010). This research revealed a close association between the invasive dove and landscapes that have been highly modified by human activities.

In addition to research coming directly from the FeederWatch database, the network of FeederWatch volunteers has proved to be an excellent source from which to recruit participants for more focused studies. For instance, the spread of a novel strain of the bacterium *Mycoplasma gallisepticum* in House Finch populations was tracked by participants in FeederWatch along with a new project (The House Finch Disease Survey) that drew from the FeederWatch community and built on the existing data-entry platform. This collaboration led to a great amount of scientific output featuring FeederWatch data (Dhondt et al. 1998; Fischer et al. 1997; Hartup et al. 2001; Hartup et al. 1998; Hochachka and Dhondt 2000) and quantified the impact of the disease on House Finch populations (Hochachka and Dhondt 2000). This research definitively demonstrated that the disease caused population declines in House Finches—a species that had previously been among the most familiar backyard birds in many locations.

Analysis of the educational outcomes of FeederWatch participation has been limited and remains a rich area for future research. Studies of the knowledge base of participants prior to engagement, and the learning achieved by FeederWatch participants, did not begin until 2006 (Bonney 2008). Although the FeederWatch audience is highly educated (76% hold a college degree and 35% hold postgraduate degrees), a survey with 2157 respondents indicated that 15% had been watching birds for less than five years and 53% self-identified as "intermediate" birders, suggesting that many participants have much to learn about birds. Indeed, one new participant who did not understand patterns of seasonal molt in birds wrote, "Who knew goldfinches could turn from dull green in winter to bright yellow in spring?"

Half of respondents to a 2006 survey reported discovering a greater diversity of birds in their yards through their participation in FeederWatch. Three in four participants reported observing interesting bird behaviors, and 70% reported learning more about seasonal changes in bird communities. Fewer than 6% of respondents failed to learn anything during their involvement, suggesting that the educational goals of the project are, at least in part, being met.

Sustainability

Sustaining citizen science programming requires a business model, particularly when there is a desire to collect data over the long term. Unlike many large-scale citizen science projects, FeederWatch does not rely on grant funding. Rather, participants fund the bulk of the program through annual fees and donations.

As with many citizen science projects, participant recruitment is a perpetual challenge. Despite a remarkable annual retention rate of greater than 70%, approximately 3000 to 4000 new participants need to be recruited annually to maintain participant numbers and widespread geographic coverage. Outreach to underserved audiences and underrepresented geographic areas (e.g., the Southwest) remains a challenge. Nevertheless, FeederWatch is an example of a sustainable and successful citizen science project that engages thousands of participants annually in productive scientific research.

🦋

FeederWatch has thrived for more than two decades as a continental-scale citizen science project that enlists eager participants to do what they were already doing—watching backyard birds—in a more scientifically rigorous way. FeederWatch is unique in that its funding is derived almost entirely from the fees paid and gifts provided by the participants. Significant support is provided to participants by project staff, but social networking tools are now helping participants assist and support one another. More than a dozen peer-reviewed scientific publications have made use of FeederWatch data, demonstrating that useful information is indeed being gathered. Great potential remains, however, for FeederWatch to contribute to population biology, community ecology, biogeography, climate- and habitat-change research, and other areas of ecological and educational research.

Acknowledgments
Special thanks to the thousands of citizen scientists who make Project Feeder-Watch possible. The support of numerous staff members at the Cornell Lab of Ornithology and Bird Studies Canada is critical to the ongoing success and development of the program. The efforts of Rick Bonney, Janis Dickinson, Erica Dunn, Wesley Hochachka, Anne Marie Johnson, and Genna Knight are especially appreciated.

Monitoring Monarchs: Citizen Science and a Charismatic Insect

KAREN S. OBERHAUSER

The Monarch Larva Monitoring Project (MLMP) is a citizen science project run by researchers, staff, and students at the University of Minnesota. It involves volunteers from across the United States and Canada in monarch research, and was developed to collect long-term data on immature (egg and larva) monarchs and milkweed habitat. The overarching scientific goal of the project is to better understand monarch distribution and abundance during the breeding season. The equally important educational goal of the MLMP is to provide citizens with hands-on experience in scientific research. Through their engagement in the MLMP, volunteer monitors enhance their appreciation and understanding of monarchs, monarch habitat, and the scientific process in general.

Project Design

MLMP volunteers include adults monitoring on their own, in organized groups, or with children. Volunteers learn about the project from butterfly and monarch Listservs, local nature centers, educational programs, friends and family, or organizations such as state Master Naturalist programs.

The central protocols for the MLMP consist of six data-collection activities designed by the scientists who developed the project, Michelle Prysby, Sonia Altizer, and Karen Oberhauser. Data-entry forms, data-display tools, and directions for participation are posted on the project website (www.mlmp.org). Participants conduct two activities once per year: (1) compose a site description, and (2) measure milkweed density on the site. In addition to these annual activities, participants (3) conduct weekly surveys of monarchs and milkweeds. They randomly choose plants to census, or on smaller sites they census all plants, and then record the number of milkweeds examined and the numbers of eggs and larvae observed, identifying larvae to instar (monarchs go through five larval instars, or stages, between molts). There are three additional, but optional, data-collection activities: (4) compare plants occupied by monarchs to random plants (height, flowering status, age, herbivore damage, and the presence or absence of invertebrates); (5) measure rates of parasitism by parasitoid flies by rearing larvae collected from their sites and reporting their fate (turned into a healthy adult, parasitized, or died of another cause); and (6) collect rainfall data. Volunteers enter all data into an online database. At the end of the season they mail hard copies of their data sheets, which project staff use to spot-check the online data and then archive.

Volunteers choose their own sites, which include backyard gardens, abandoned fields, pastures, and restored prairies located throughout the monarch's breeding range (mainly the eastern half of the United States and southeastern Canada). The only requirement is that the sites contain milkweed, the host plant for larval monarchs.

MLMP participants comprise all ages and demographics. As of early 2011, over 500 primary individuals had been associated with sites, but the number of volunteers is much larger because, until 2008, only one person could register as a volunteer for each site, and many volunteers monitor in groups. For example, almost half of participants report that they monitor monarchs with up to twenty-five children.

Over 55% of the registered individuals have monitored in multiple years, and from one year to the next the project has an average retention rate

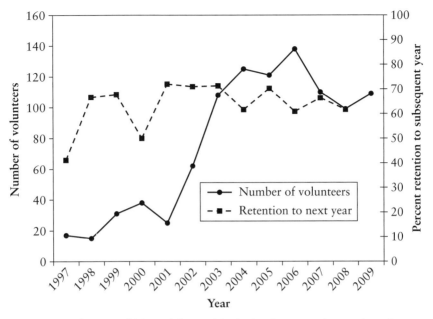

Figure 2.3. Total registered Monarch Larvae Monitoring Program volunteers in each year (actual numbers are higher owing to multiple volunteers per site) and retention to the next year (percent of volunteers in a given year that also monitored the following year).

of 67% (Figure 2.3). Many individuals monitor more than one site; by 2009, over 700 sites in 28 U.S. states and 3 Canadian provinces had been monitored.

Although this project uses specified protocols, many participants, especially teachers, engage youth participants in a variety of related activities of their own design, including defining new questions for study, developing hypotheses, designing data-collection methodologies, interpreting data, and drawing conclusions. In one case, an independent investigation instigated by a volunteer led to a peer-reviewed paper that included three volunteers among its authors (Oberhauser et al. 2007). Many independent projects are highlighted in the project's annual newsletter.

Participant Interaction

The level of interaction with participants is variable depending on their needs and desires. Some volunteers send frequent e-mails with updates, observations, and questions. Project leaders try to respond within a day, and often use volunteer communications in the annual newsletter.

Summaries of project data are available online and in the newsletter, which also includes articles of general interest on monarch biology and conservation, monitoring reminders and recommendations, and information on MLMP support materials. Raw data from individual sites can be downloaded from the website, and other raw data are made available on request to participants, the media, or scientists for informational, scientific, and educational purposes.

It is clear from their communications that volunteers are excited about and learning from their experiences, and that they feel a connection to project personnel:

> Hi, Karen. You will be as thrilled as I [am]. . . . [We] had 3.5 inches of rain from May 1–4 . . . so I waited anxiously til May 7th to check my gardens. Hold on . . . here are the results of that day's monitoring: 121 plants checked, 17 eggs, 95!!!!!! 1st stage instars, 4 2nd instars and 1 female adult. . . . I'm delighted to give this report to you.

Many find opportunities to share project findings with others, often for conservation purposes. For example, one volunteer was asked by a National Wildlife Refuge biologist for background information for a grassland management plan. The biologist wanted to plan a mowing regime around monarch breeding, and the volunteer was able to share findings from her own monitoring site and others.

Volunteer communications about patterns and other natural phenomena often illustrate sophisticated scientific thinking:

> This year I have been taking wing measurements of the monarchs in the parasite study. I anticipate seeing a decline in wing size as the fall progresses because the food will be dying back. Is there a protocol in place for this? I close them up and measure from the large white spot to the tip of the forewing. For comparison this should work well as long as I stay with this method for this year, but I would like to do better next year if possible.
>
> . . . Do you think that somehow monarchs can transfer information about where they were born and pass it along to future generations so that each year more will return to lay eggs? If they are born with the knowledge of how and where to migrate to in winter, even having never been there before, is it possible that they could also pass along this other information?

Volunteers are engaged in a variety of formal and informal teaching and mentoring activities. Many speak to school groups about their research or give presentations for organizations such as Project Wild or state Master Naturalist programs.

Training and Educational Resources

MLMP participants learn project procedures by reading directions online or by attending in-person trainings at which they learn about monarch biology, practice monitoring and data-entry protocols, and receive needed materials. From 2001–2004 a National Science Foundation grant supported train-the-trainer workshops throughout the eastern United States; during these workshops, almost 200 naturalists, other informal science education personnel, and teachers became MLMP trainers. Many of these individuals still conduct local and regional training sessions and organize monitoring teams at their sites, and long-term volunteers can request to be added to the roster of trainers. All MLMP training workshops are announced on the project website. Additionally, individuals contact project leaders for help with specific protocols or to find out how to get started on the project.

A variety of MLMP training and educational resources support the program. The most-used resource is a monitoring kit including a clipboard with directions printed on the back, a hand lens, a foldable meter stick, pencils, a rain gauge, and a field guide developed specifically for the project to help participants identify and describe invertebrates found on milkweed plants (Rea et al. 2002). All of these materials come in the pocket of a field apron decorated with the MLMP logo. Volunteers can also download an interpretative sign, available as both a modifiable PowerPoint file and a PDF file, to post at their sites to inform the public about monarchs, the MLMP, and their site.

Data Collection and Validation

Inaccurate data are most likely to arise from either identification/observation errors (e.g., a volunteer mistakes milkweed latex for an egg or misses a first instar larva on a plant) or from nonrandom observations (e.g., a volunteer looks only at milkweed plants that are in good condition or includes plants that are outside the transect during milkweed density counts). These errors can result in two undesirable outcomes: Web visitors may view erroneous data (for data sets, such as monarch density, which are posted), and project staff or others may use erroneous data in publications.

Project staff lessen the chances of receiving inaccurate data through training, hands-on practice whenever possible, easy-to-follow instructions, and by stressing the importance of random sampling. They check all data before using them in publications, and delete individual data sets if there is good reason to believe that they are erroneous or incomplete. For example, some MLMP volunteers see no eggs but many larvae. The presence of larvae in the absence of eggs is impossible, because the high incidence of predators in the wild means that eggs far outnumber larvae at field sites. When project data are being used to estimate egg-to-larva survival, data sets such as these are omitted. These data are useful, however, in assessing the presence of monarchs in a given place at a given time, as repeated observations for occupancy modeling (see Chapter 7). Volunteers who send questionable data sets are interviewed to determine whether they have data-collection problems that can easily be addressed, but in some cases the volunteers simply seem unable to see monarch eggs and early instar larvae. Their data sets are not used in analyses, but project staff have chosen not to hurt the volunteers' feelings, so the data are left on the website. Thus, people looking at individual site data may occasionally view erroneous data. Project staff encourage other researchers to consult them before using any downloadable project data, and they explain potential problems with the data to researchers who request data that are not downloadable from the website.

In sum, project leaders anticipate common problems and try to prevent them, but also acknowledge and deal with problems during data analysis.

Impacts

Educational goals of the MLMP have been evaluated both anecdotally and in formal surveys and interviews. Whereas evaluations have focused on the outcomes for youth, communications from adult volunteers as described earlier demonstrate that many adults have learned a great deal about the process of science through this project. Anecdotal evidence suggests that youth volunteers have learned about social, process, and content aspects of science. Many mention the importance of collaboration: "It is amazing to me that when people all over the country take a little time every week, and even more in some cases, to count butterfly eggs, the end result is a network of data that can help us decipher where the butterflies go and when and how. . . . This is real life proof that when everybody works together, things can be done." It has also helped teach youth to be scientists: "The MLMP has helped me become a better scientist in so many ways. Most importantly, it gave me a large interest in science. It encouraged me to ask questions such as why and how and to find these answers through experimenting."

Kountoupes and Oberhauser (2008) summarize a formal evaluation of project impacts on youth that was conducted through surveys and interviews of adult mentors (family members, teachers, and informal science education [ISE] professionals and volunteers). This mixed-method approach elucidated the contexts, outcomes, and promising practices for engaging youth in the MLMP. Adult volunteers reported that youth were successful at most of the project activities. The adults made innovations to increase the success and educational value of the project for children, and were careful to ensure that these innovations did not compromise data integrity. They felt that the children gained an understanding of real scientific research, and that they were proud of their contributions. Many teachers who monitored monarchs with youth groups during the summer highlighted social aspects of the program. One interviewee described the experience for her group of children as "science bonding," and another said that the project provided an alternative to sports-centered recreation.

From a scientific perspective, MLMP data have been used in several publications by both project personnel and other scientists. These papers have

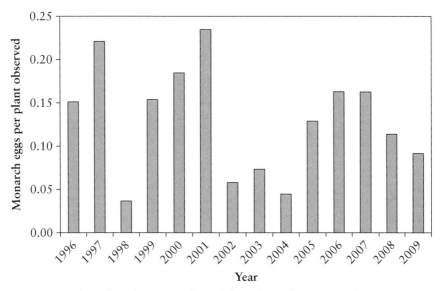

Figure 2.4. Peak number of eggs per milkweed plant observed in Upper Midwest (MN, WI, MI, IA) monitoring sites. This number reflects the size of the monarch population that will migrate to Mexico in a given year. The mean across all years is 0.13 eggs per plant.

addressed impacts of crowding on disease incidence in monarchs (Lindsey et al. 2009), potential impacts of climate change on monarchs (Batalden et al. 2007), incidence of a fly parasitoid in monarchs (Oberhauser et al. 2007), natural enemies of monarchs (Prysby 2004), temporal and spatial patterns in monarch abundance (Prysby and Oberhauser 1999, 2004), and impacts of Bt corn pollen on monarchs (Oberhauser et al. 2001). Year-to-year variation in monarch density is useful for assessing long-term trends in monarch population numbers (see Figure 2.4). Additionally, MLMP coordinators are cooperating with coordinators of other monarch monitoring programs to develop a shared database that will encourage collaborative data analyses, thus leading to better understanding of monarch population dynamics (www.monarchnet.org).

Sustainability

The MLMP requires both financial and human resources to continue. A grant from the Xerces Society funded early training workshops, and an NSF grant (ISE 0104600) funded training, materials development, and database development activities three years after the project started and was largely responsible for its early growth. Since that funding ended, the project has been maintained without external support devoted exclusively to the MLMP, although educational projects that are outgrowths of the MLMP have been funded by the National Science Foundation (ISE 0917450), the

Medtronic Foundation, and the Minnesota Office of Higher Education. Personnel who are supported by the University of Minnesota Extension program communicate with volunteers, write the annual newsletter, talk about the program in a variety of venues, analyze the data, and conduct their own monitoring. Resources for program management and marketing are critical to project growth (see Chapter 4), and it is possible that lack of resources accounts for the leveling off of volunteer numbers since 2005 (Figure 2.3). Currently approximately one FTE is devoted to the MLMP, but this effort varies seasonally and does not include time spent on data analysis.

External support for the MLMP has always had an education focus. While there are clear educational benefits of the project, and project personnel have a strong engagement with science and environmental education programs (see www.monarchlab.org), this focus has meant that some of the scientific possibilities have not been addressed. In 2010 and 2011, however, collaborations with other monarch monitoring programs (see www.monarchnet.org) are resulting in exciting analytical progress.

🦋

The MLMP has resulted in a database that has helped scientists, policy-makers, citizens, and the media appreciate the importance of long-term data in understanding animal, especially insect, population trends. From a conservation perspective, the protocol and findings were utilized in the North American Monarch Conservation Plan (Commission for Environmental Cooperation 2008), and citizen volunteers have reported both an increased awareness of conservation issues and increased local conservation activism in responses to program surveys (Kountoupes and Oberhauser, unpublished data). Additionally, nature center professionals, teachers, and parents have used the program as a science education tool with youth, and report increased familiarity with the process of science as a result.

Acknowledgments

I thank the hundreds of MLMP volunteers who have collected data for this project, especially Pete and Sanny Oberhauser, who have monitored for more years than any other volunteers, and Cindy Petersen, who has engaged well over 200 middle school students in this project. Without this dedicated group, our knowledge of monarch biology would be much poorer. Sonia Altizer, Michelle Prysby, and Michelle Solensky were instrumental in starting the MLMP, and Liz Goehring, Beth LaVoie, Reba Batalden, Alma DeAnda, Grant Bowers, Jolene Lushine, and Matt Kaiser have made important contributions. The National Science Foundation (ISE 0104600 and ISE 0917450) has supported project development and dissemination, and the Xerces Society supported early project dissemination.

Neighborhood Nestwatch: Mentoring Citizens in the Urban Matrix

PETER P. MARRA AND ROBERT REITSMA

In our rapidly changing world, nearly 1 million acres in the United States are converted into urbanized landscapes each year (Heimlich and Anderson 2001). The continuing trend of habitat destruction for urban growth represents the most serious threat facing animal populations in North America today. Ironically, we know little about the biology of most wildlife living in urbanized areas (Marzluff et al. 2001). Urban wildlife can persist either in remnant forest fragments or within the remaining matrix of yards, businesses, and other forms of development. Therefore, research within the matrix is critically needed to uncover the fundamental and applied problems facing urban wildlife and to provide information about what measures might mitigate such threats.

Equally important is increasing awareness and understanding of urban wildlife needs and promoting a conservation ethic among urban residents, many of whom have become detached from their surrounding environment. Such detachment results in human insensitivity to natural processes and a lack of understanding of the need for environmental protection. Changing people's perceptions about the importance of wildlife habitat in the urban matrix is the first and most challenging step toward mitigating human impacts.

To meet this dual challenge, in 2000 the Smithsonian Institution created a citizen science project called Neighborhood Nestwatch (NN). Focusing on birds breeding in the backyards of citizen volunteers in the Washington, DC–Baltimore, Maryland, metro region, NN—like most citizen science projects—has objectives in both science and education. Its scientific goal is to learn how urbanization affects the reproduction and survival of common backyard birds. On the education front, the program brings Smithsonian scientists face-to-face with volunteers in their backyards where the scientists teach the participants about birds and the process of conducting research. Although the approach of personally mentoring participants is labor-intensive, it promotes active relationships between scientists, citizens, their backyard wildlife, and the surrounding environment. The end result is that scientists gain significant information about urban birds, while participants engage deeply in research and learn about the role that their property plays in overall ecosystem health. In sum, we turn the backyard of the citizen into a study site for the scientist and a classroom for the participant.

Project Design

As mentors and research biologists, Smithsonian staff members visit each volunteer citizen scientist in their yard at least once annually during the spring or early summer to carry out a number of research projects and educational activities. The volunteers live within a 60 km radius of the Washington, DC, area and come from all walks of life.

Project activities center on eight focal bird species that may nest on the participants' property: Carolina Chickadee, House Wren, Carolina Wren, American Robin, Gray Catbird, Northern Mockingbird, Northern Cardinal, and Song Sparrow. We chose these species because they commonly occur across the entire region in urban, suburban, and rural backyards, allowing us to examine the effects of land use on their biology. In addition, they are relatively easy to capture and not too difficult for novices to observe. On each visit, a Smithsonian scientist teams up with the NN participants to set up mist nets and play recordings of bird songs to lure birds into the nets. Once birds are in the hand, the data collection begins. The scientist handles the birds and bands them with a unique combination of two color leg bands and a United States Fish and Wildlife Service band. The scientist also takes measurements including wing, tarsus, and bill length, as well as body mass and feather and blood samples. The participant typically observes the process and records the data.

Each measurement can provide unique insight into the lives of birds that persist in the urban environment. For example, by comparing measurements of body size to weight, we can learn about the condition of the bird. Feather and blood samples provide details on genetic composition as well as contaminant and disease exposure. The unique combination of color bands allows the citizen scientists to continue to recognize and observe "their" individual birds throughout the year and, with luck, over several years. Data provided by such resightings can provide tremendous insights into annual survival. For example, in the case of a migratory gray catbird color banded in the backyard of Joan W. of Fairfax, Virginia, the return of the same bird (red band over blue band on the left leg and aluminum band on the right) over three subsequent breeding seasons introduced this participant not only to the notion that her bird survived for three years but also to the fact that, when her bird disappeared from her backyard, it migrated to the tropics where it may have been foraging next to a jaguar in Belize or behind a Mayan hut in Guatemala.

Scientists and citizen participants also engage in a variety of other activities. These include searching for and monitoring nests at regular three-day intervals during both the incubation and nestling periods, which results in valuable data on clutch size, rates of nest predation, and number of young fledged—all information essential for estimating reproductive success. Participants are also asked to collect data on the height of the nest, the plant species in which the nest is located, and the number of outdoor cats and bird feeders in the yard. In addition, Smithsonian scientists conduct annual bird censuses in each of the backyards to determine the relative abundances of birds across the urban matrix. Repeating censuses over multiple

Figure 2.5. Neighborhood Nestwatch founder Peter Marra measures and bands a Northern Cardinal in the backyard of a Takoma Park, Maryland, family with homeowners and neighbors looking on.

years provides essential information on population change over time and across space.

Participant Interaction, Training, and Educational Resources

The interactive nature of the visits creates active engagement between research staff and participant. The scientist explains field techniques and general bird natural history, suggests enhancements to improve backyard habitats for birds and other wildlife, and provides direction to the participants on how to record data, search for nests, and resight color-banded birds. After the visit the newly graduated citizen scientists begin collecting data independently. In general, the entire NN experience offers "on-site" training that makes a connection at a very personal level, fostering a greater understanding of urban ecosystems and changing people's perceptions of their backyard habitats. Moreover, participants often invite neighbors to be part of the scientists' visit. In this way the interaction between scientists and citizens becomes a community event that expands NN outreach to the broader region.

The NN program offers additional guidance through handouts with hints for nest searching, bird observation, and the resighting of color-banded birds. An annual newsletter disseminates information on backyard biology themes, general project updates, results from prior seasons, descriptions of the publications that have come from NN research, and information on where NN staff have gone during previous years. Throughout the year NN staff stay in touch with participants by phone and e-mail to resolve a variety of issues ranging from general bird natural history to technical help with data entry. Finally, the website (nationalzoo.si.edu/goto/nestwatch) includes detailed text that provides help with data collection as well as other information such as a listing of backyard habitat enhancements. Collectively, through communication in and out of the field, NN creates a personal relationship between scientists and citizen scientists, which ultimately fosters a productive, effective, and exciting research program.

Participants are always encouraged to submit suggestions for improvement. Suggestions vary, but typically involve requests for more information on how to find nests and resight birds and for more online training or help with data entry. All requests are answered either by e-mail or phone and serve as motivation to better achieve project goals.

Data Collection and Validation

Although the geographic scope of NN is not as broad as other citizen science projects discussed in this book that span the United States and populated parts of Canada, the program has impressive depth in terms of the type and quality of data that are collected at each NN site by both participants and Smithsonian scientists. Participants record observations on provided data sheets and either mail the forms to the Smithsonian or enter their data on the project website. Web data are converted to an e-mail message that is sent to a staff member who checks for errors before the data are analyzed. Project staff enter banding, bird censusing, and vegetation data into this same database.

Impacts

Nestwatch has been effective in reaching out to both scientific and lay communities. To date, twelve peer-reviewed journal articles have been published using data collected by both participants and scientists. The papers range from studies of reproductive success of the eight NN species (Newhouse et al. 2008; Ryder et al. 2010) to studies of disease dynamics (Griffing et al. 2007; LaDeau et al. 2007), contaminant loads (e.g., lead) in birds (Roux and Marra 2007), and impacts of urban noise on song production (Dowling et al., in revision). To demonstrate the effectiveness and importance of the NN approach to scientific research, consider the following example, in which we quantified bird reproductive success, an essential process for determining habitat quality and for understanding how populations persist.

Using data generated by NN citizens and scientists for five of our target species, we estimated the length of time that nests with eggs and young persisted before the young fledged or were eaten by predators (i.e., nest survival). We conducted the study using a sample of more than 400 nests located by an annual participant base averaging 170 observers. These nests were observed both by participants and scientists from within the matrix along our urbanization gradient (Ryder et al. 2010). We found that nest survival (at the egg and nestling stages) was higher in urban regions than in forested areas (Figure 2.6). We had expected the opposite result, because predators such as crows and cats tend to be more numerous in urban areas. Instead, our results suggest that the more diverse predator communities in forested environments are responsible for higher rates of nest predation there. Nest success increases in more developed landscapes apparently because roads and other forms of urban disturbance cause the extirpation of natural nest predators such as snakes and small mammals.

These trends were supported by an additional experiment carried out by a subset of NN participants who were each provided with an artificial nest and two quail eggs. They were instructed to put the eggs inside the nest and to place the nest outside in a location that mimicked the natural nest site of a catbird or cardinal. Finally, they were asked to check the nest twelve days after placing it to determine if the nest and its eggs had survived. The data from this independent experiment mirrored the results from our natural experiment. Finally, and perhaps most exciting for the field of citizen sci-

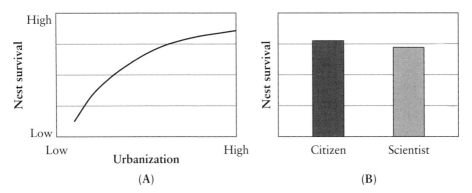

Figure 2.6. (A) Nest survival of five species of common backyard birds increases with greater urbanization as measured by the percentage of impervious surface area at 500 m radii around participants' homes (B) Nest survival estimates for five species of common backyard birds collected by scientists and Neighborhood Nestwatch participants did not differ, suggesting that citizens can collect data equal in quality to those collected by trained scientists.

ence, we contrasted data collected by citizen scientists with those collected by Smithsonian scientists and showed them to be of comparable quality (Figure 2.6 B). Collectively, these results are important because they provide the first substantial data on reproductive success of breeding birds in the urban matrix. They also demonstrate that amateur scientists, if given proper mentoring and tools, are capable of providing high-quality data.

Assessing the effectiveness of NN at reaching the public with our education message has been a priority since project inception. Although our participant base is small compared with some projects, indications are that the impact on participants is relatively large. In 2001, a study using pre- and post-visit surveys of forty-five NN participants showed that 87% of participants increased their knowledge of biological attributes such as territoriality, duration of nesting cycle, habitat preferences, behavioral differences among species, and predator ecology. In addition, 83% of participants gained greater awareness of their local environment as evidenced by increased attentiveness to the presence or absence of backyard birds and whether their property provided adequate bird habitat. Finally, 56% of participants enhanced their yards for nesting birds by providing nest boxes, planting bird-friendly vegetation for shelter and food, leaving beneficial plants undisturbed, and keeping cats indoors (Evans et al. 2005).

In another assessment conducted jointly by the Cornell Lab of Ornithology and the Smithsonian Migratory Bird Center in 2007 and 2008, NN participants were found to be more proactive with their research activities relative to people in other types of citizen science projects focused on nesting birds. More specifically, when compared to projects that were entirely Internet based or those that were Internet based but included a single group workshop or orientation session before the breeding season, NN participants were more likely to monitor nests, to record and submit data about nests, and to ask questions regarding nest monitoring (Table 2.1). This assessment also showed that although NN participants enter the project with a relatively low knowledge base of bird natural history, their experience in the project led to gains in knowledge of nest monitoring practices, biological vocabulary, and variation among species (Bonney et al., in prep). We attribute these patterns to direct contact between participant and scientist during backyard visits and to frequent e-mail correspondence.

Finally, the NN program has provided enormous professional training and research opportunities. To date, approximately forty undergraduate interns, two high school students, two masters- and doctoral-level students, and three postdoctoral fellows have interacted with the participant base and conducted research using NN sites or participants or both. Nestwatch staff have partnered with high schools and local universities to incorporate formal courses of study using case studies of NN research as the structured curriculum. Such research questions about the impacts of urbanization

TABLE 2.1.
Percentage of citizen scientists participating in three different projects focused
on nesting birds

	Citizen science program		
Program activity	Neighborhood Nestwatch	NestQuest[a]	NestWatch[b]
Monitor nests	98	74	93
Record data	98	62	88
Submit data	98	62	83
Ask questions	98	54	69

Note: The data are based on surveys of new participants conducted after the end of the 2007 breeding season.
   [a] NestQuest: Groups of citizens participate in single daylong workshops conducted by a Cornell University biologist before engaging in project activities in their own yards.
   [b] NestWatch: A citizen science program maintained by the Cornell Lab of Ornithology that is entirely Internet based, from initial instruction to participant data entry.

and student opportunities for training are essential for understanding and training the next generation of scientists to work in the emerging frontier of urban ecology.

Sustainability

Nestwatch has experienced rapid growth from 45 backyard sites in the inaugural spring of 2000 to 281 sites in 2009. Each site or property contains between one and six participants, so the program now reaches close to 1000 people annually. Recruitment has occurred primarily through public lectures and stories in local print media, beginning in 1999 when we gave concept talks at local Audubon chapters and bird clubs and placed advertisements in many Audubon newsletters. The initial response was overwhelming, and ten years later, requests continue to pour into our offices.

Because the program is labor-intensive and thus costly (e.g., NN can cost up to $42,000 per year including staff salaries), expansion is challenging. One approach is to use established science centers at key geographic localities or hubs and to capitalize on the huge numbers of volunteer bird banders who are eager to catch birds and communicate their excitement about science. Another idea is to develop collaborations with other scientists who are conducting similar research across the United States. These efforts would have similar broad scientific goals, but each site could have region-specific research questions.

For the moment we have stopped all recruiting efforts and have created a waiting list that has a minimum of fifty citizens each year. Wait-listed people are encouraged to join the Cornell Lab of Ornithology's NestWatch program, for which participants across the nation find and collect data on nesting birds on their own without being tutored during an encounter with a scientist.

♥♥

Neighborhood Nestwatch uses a citizen science approach not just to gather important scientific data on birds but also to give people a closer glimpse of wildlife found in their own backyards. By actively involving participants in the collection of scientific data, NN brings people a greater understanding of wildlife biology and the threats posed by urbanization. Formal education and traditional scientific approaches are not enough to ensure scientific literacy of citizens in a world where ideas, technology, and the landscape surrounding us are constantly changing. Making scientists and their science more accessible and approachable puts NN in a position to significantly impact the public's understanding of science.

Acknowledgments
The results of Neighborhood Nestwatch are possible thanks to the tireless efforts of our citizen participants and our committed staff. Funding has been provided by the National Science Foundation, the Wallace Genetic Foundation, the Mars family, and the Smithsonian Institution.

Project BudBurst: Citizen Science for All Seasons

SANDRA HENDERSON, DENNIS L. WARD, KIRSTEN K. MEYMARIS, PAUL ALABACK, AND KAYRI HAVENS

Project BudBurst (PBB; www.budburst.org) is a national field campaign designed to engage the public in collecting data about plant phenology. Participants make careful observations of phenological events such as emergence of first leaves, first flowers, and first ripe fruit of a diversity of native and nonnative plant species, including trees, shrubs, flowers, herbs, weeds, and ornamentals. Compared to other projects described in this book, PBB is a newcomer in the world of citizen science and environmental research. However, from its initial ten-week pilot in the spring of 2007 through the completion of its most recent season in 2010, PBB has garnered significant interest and attention from the public and the media, emerging as a viable and engaging plant phenology citizen science project. In addition, it is starting to see results.

Designed to collect continental-scale phenological data, the goals of the project are to increase awareness of the effects of climate and environmental change on plants at a local level; increase understanding and appreciation of science by engaging participants in the scientific process; and encourage individuals to spend time outdoors appreciating nature, especially plants.

The underlying mission of PBB is simple—to get people from all walks of life to be more aware of the natural world around them and to record observations that advance the scientific understanding of changing environments. While it is safe to assume that most individuals recognize seasonal changes in their environment (e.g., trees beginning to leaf out, flowers first appearing in spring, fruits ripening in the summer, and leaves falling off trees in the fall), it is less likely that they are familiar with annual or decadal changes in the timing of these events or of phenology as an area of scientific study. Observing plants is a relatively easy way to get people in all areas of the country involved in environmental research. Helping participants of all ages learn to make accurate observations of plant phenophases is relatively straightforward and does not require any special instruments.

Plant phenologists have long been interested in how changes in the environment can influence the timing of specific life-cycle events. Plants use cues in the environment such as variation in day length, temperature, and precipitation to determine when to put out new leaves, open flowers, or ripen fruits. Because many such phenomena are sensitive to small variations in climate, especially temperature, existing phenological records are used as "proxy" data to better understand past climates for periods or locations where no instrumental records are available (Miller-Rushing and Primack 2008; Primack, Higuchi, et al. 2009; Primack, Miller-Rushing, et al. 2009; Schwartz et al. 2006).

The Chinese are thought to have kept the first written phenological records dating back to about 974 BCE. Many individuals, including some famous observers (e.g., Carolus Linnaeus, Thomas Jefferson, Henry David Thoreau, and Aldo Leopold), have kept diaries and garden journals that included the dates of phenological events such as budburst, first flowering, and leaf fall, along with recorded prevailing weather conditions. Thoreau noted hundreds of plant species in the Concord, Massachusetts, area in his journals. The growing scientific interest in plant phenology has reinvigorated interest in keeping these kinds of detailed records.

Diaries and journals are an excellent example of the usefulness of phenology data when consistently collected over a long period of time. In the early twenty-first century, Miller-Rushing and Primack (2008) built on Thoreau's work. They collected contemporary phenological observations on hundreds of plant species that were carefully tracked by Thoreau over 150 years ago. Comparisons between the historical and modern data were used to demonstrate profound changes in plant phenology at Walden Pond, with timing of spring events advancing by an average of one week for many species. Further studies of these data have provided support for the novel hypothesis that species that can quickly change their phenology in response to changes in climate tend to be more common today than those that have not significantly changed their phenology since the time of Thoreau

(Primack, Miller-Rushing, et al. 2009). Data such as those being collected in PBB will provide a much broader geographic test of this hypothesis with direct implications for biodiversity conservation.

A similar project that began in the 1950s, The Lilac Network, has already made significant contributions to the science of phenology through observations made by citizen scientists from around the country (Schwartz et al. 2006). Analysis of resulting data indicates that spring is now arriving five to six days earlier in some areas. The success of The Lilac Network helped launch the USA National Phenology Network (USA-NPN), which was established in 2004 to monitor the influence of climate on the phenology of plants, animals, and landscapes. This national network has further generated scientific interest in plant phenology in the United States and in providing data and models that may help scientists monitor and predict drought, wildfire risk, biological invasions, and the spread of diseases more accurately than currently possible. Thousands of individuals across the country have contributed observations to The Lilac Network and USA-NPN.

Project Design

In its first years of operation, USA-NPN focused on developing protocols, tools, and programs targeted at researchers, graduate students, and agency scientists, all of whom are interested in applications of these data to the development of predictive models. PBB grew out of early planning meetings for USA-NPN and continues to be an important partner in the network. It focuses only on plant phenology and attempts to engage a broad audience of individuals who may or may not have background or experience in making observations of phenological change. An important objective of PBB is to test and validate broader monitoring strategies. Initially we have cast our net widely with the intent of offering more selective methodologies over time for both formal (K–12) and informal audiences.

In the fall of 2006, PBB secured a small amount of funding to launch a pilot project, which was planned and implemented by science educators and scientists from several organizations and institutions. The timing of PBB's pilot in the spring of 2007 benefited from increasing public interest in climate change coupled with the widespread accessibility of the Internet.

PBB participants register on the website and provide information on site location(s), habitat characteristics, and identification of species to be monitored. Participants then make daily or weekly observations when plants get close to reaching a particular phenophase (e.g., first leaf unfolded, first flower, end of flowering, first ripe fruit, and when 50% of branches have fallen leaves), recording the day of the event. Volunteers then submit their data directly to the website.

Participant Interaction, Training, and Educational Resources

To support the 2007 pilot project, identification guides were developed for sixty target species chosen for their broad geographic distribution, ease of identification, and scientific interest. Allowing observations of "other" plant species turned out to be an important decision, because many participants made observations of species that were not on the target list. Because of PBB's strong focus on education and outreach at a national scale, it was important that we make participation as inclusive as possible. If we did not accept non-PBB targeted species, we would exclude individuals who did not have ready access to our target species. To encourage and support observation of specific plants deemed useful for scientific analysis, we developed America's Ten Most Wanted plant list.

All materials necessary to participate in PBB were made available at no cost on the website. The online reporting interface was designed to be user-friendly for participants of all ages. Recruitment was an overwhelming success because of the immediate response and high level of interest from volunteers and the media. This success attracted additional funding from more agencies and institutions to support a yearlong enhanced project in 2008 in which thousands of citizen scientists participated. Fifteen additional plant species were added to the core species list to better represent southern and western regions, increasing the total from sixty to seventy-five species.

In 2008, nearly 5000 observations of phenophases were reported online from participants in forty-nine states, providing useful baseline data. Participants included educators and students from elementary to graduate school, demonstrating the versatility in implementing PBB in formal educational settings. Other participants included amateur naturalists, gardeners, visitors to botanic gardens and nature centers, and after-school programs.

In 2008, the species most often observed were the common lilac and the common dandelion. Together, these two easily identifiable and geographically widespread plant species made up 13% of all reported data. In total, however, data were received on more than 500 additional species or varieties. The earlier phenophases (first leaf, full leaf, first flower, full flower) had the most observations reported. In 2009 some autumnal events (e.g., leaf color change, leaf drop) were added. The distribution of phenophase observations for 2008–2010 are shown in Figure 2.7.

Data Collection and Validation

PBB staff made several changes to the project based on participant feedback. For example, it became clear that users need to be able to see and review their own data and that the scientists working with PBB need to be able to associate observations with sites and observers. For these reasons a

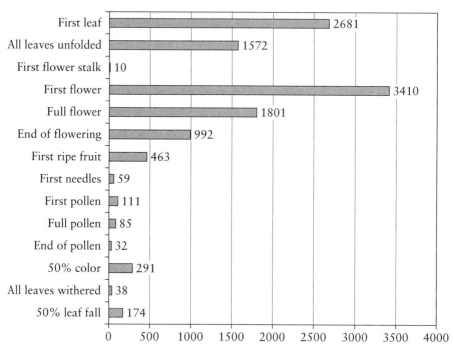

Figure 2.7. Frequency distribution of phenophase observations collected during the first four years of Project BudBurst.

feature called "My BudBurst Data" was added. Also based on participant feedback, PBB developed illustrated phenophase field guides for the seventy-five core species. Phenophase names and definitions were modified to be compatible with USA-NPN, and better descriptions of the phenological development for particular species groups were provided. Additional features were added to the website, including a weekly blog during the growing season and a "plant of the week" page. To provide immediate feedback for participants when they submitted their observations, live mapping was added that showed the locations of the last 100 observations submitted to the website, including information on species and phenophase.

By the end of the 2010 field campaign, more than 12,500 observations had been submitted to the Project Budburst database. The most commonly reported species were the common lilac and forsythia. Other top species included common dandelion, flowering dogwood, red maple, red osier dogwood, and chokecherry.

Impacts

Early analysis of the data hints at its potential utility. A plant scientist at the Chicago Botanic Garden compared 2007–2009 first flower observations from Chicago with observations made by Swink and Wilhelm from the 1950s to 1994 and published in *Plants of the Chicago Region* (1994). Fifteen species had both PBB and historical observations, and seven of them flowered earlier in one or more of the last three years than ever observed by Swink and Wilhelm (Table 2.2). PBB scientists will continue to look for historical data sets for comparisons.

Considering the project's educational goals, Web traffic and media attention can be used as preliminary metrics for success. Since the spring of 2007, PBB has been the subject of over 200 articles, interviews, and broadcasts in major news outlets including print, radio, television, and Web. A Google search for "Project BudBurst" at the end of 2009 revealed over 32,000 entries. These measures could indicate that the goals of increased awareness of phenology and increased awareness of the impact of climate change on plants are beginning to be met.

Website traffic appears to be linked to media, publicity, and time of year. In 2007 there was almost no traffic to the website after the initial three-month campaign. In 2008 through 2010, traffic peaked each spring but then tapered off as expected. Despite the seasonal ebb and flow of website traffic, the overall number of site visits has continued to increase each year, suggesting that PBB is sustaining interest and participation beyond the heavy media attention afforded during the spring season.

TABLE 2.2.
Comparison of earliest first flowering date for seven species common to the Chicago area

| Species[a] | Earliest first flower observations | | |
	Swink and Wilhelm (1994)	Project BudBurst	Days advanced
Forsythia (*Forsythia x intermedia*)	April 25	April 1	−24
Spiderwort (*Tradescantia ohiensis*)	May 14	May 3	−11
Dogtooth violet (*Erythronium americanum*)	April 6	April 1	−5
Red maple (*Acer rubrum*)	March 20	March 6	−14
Mayapple (*Podophyllum peltatum*)	May 1	April 26	−5
Lilac (*Syringa vulgaris*)	May 3	April 16	−17
Black locust (*Robinia pseudoacacia*)	May 9	April 20	−19

Note: The historical Swink and Wilhelm (1994) data were collected from the mid-1950s through 1994, and the contemporary Project BudBurst data reflect observations made from 2007 through 2009.
[a] Common name (*Genus species*)

PBB is based, in part, on the collection of data for use by scientists who are interested in phenology and climate change. As with many citizen science projects, the appeal of participation results in part because participants believe the data being collected are useful to scientific research (Krasny and Bonney 2005). If data are not being used, project recruitment is misleading and could eventually lead to less enthusiasm and participation. For this reason, efforts have been made to enable sharing of PBB data with partners such as USA-NPN and the National Ecological Observatory Network (NEON).

Sustainability

To be successful in its goals, PBB needs to continue to provide better data quality assurance, better geographic coverage of a core set of species, export options, and visualizations of project data to enhance their usefulness for phenologists and educators alike. Future plans include providing live maps that show patterns from plant phenology observations in relation to both large-scale and localized weather patterns. We also plan to link our data to weather information to show how weather patterns can influence phenology from region to region and to illustrate how plants vary in their responses to environmental cues across the country. All of this information should help scientists develop better multiscale assessments of the ecological effects of climate change on ecosystems across the country.

In addition, Project BudBurst is developing during a time when social networking offers new ways of attracting and involving participants. Its future success depends on utilizing social and mobile technologies. UCLA's Center for Embedded Networked Sensing has partnered with PBB since 2009 to create a technology-enabled framework for participatory learning based on the use of mobile phones for data collection and reporting. Participants will be able to make observations, enter data on their phones, take photographs (useful as additional documentation and for quality assurance purposes), and automatically verify and submit their data. This has the potential to make reporting of data much faster and easier and should help engage new audiences in citizen science.

On reflection, the PBB managers were not ready for the initial overwhelming interest in the project. Responding and reacting to questions during and after the pilot in 2007 dominated the limited time available. The initial years would have been more efficient had there been adequate time for planning and evaluation. However, PBB did make it past the start-up years attracting the attention of many programs and collaborators as evidenced by the diverse funding agencies that supported the project in its early years. PBB moved to NEON as its permanent home in 2010. As a result of this move, PBB is able to give adequate attention to planning, evaluation, recruitment, and retention.

The long-term success of PBB lies in its ability to take advantage of emerging technologies, develop materials for formal and informal education, and demonstrate to researchers that project data are useful. PBB is well on its way to becoming an important citizen science project, engaging the public in research that will greatly enhance our understanding of plant phenology and, ultimately, the impacts of climate change on our environment.

Acknowledgments
The authors wish to acknowledge funding provided by the National Ecological Observatory Network, National Science Foundation, NASA, U.S. Geological Survey, U.S. Fish and Wildlife Service, National Fish and Wildlife Foundation, U.S. Bureau of Land Management, USDA USFS Southern Research Station, and National Geographic Education Foundation. Project BudBurst is comanaged by NEON and the Chicago Botanic Garden.

3

Using Bioinformatics in Citizen Science

STEVE KELLING

Advances in informatics—the gathering, management, and analysis of large quantities of data using computational techniques—are creating many opportunities for engaging volunteers in scientific study. Placing informatics tools in the hands of continental or even global networks of volunteers provides the potential to study processes that occur across very large spatial, temporal, and organizational scales. For example, a continent-wide network of bird observers participating in eBird (ebird.org) allows researchers to monitor the spread of the Eurasian Collared-Dove (*Streptopelia decaocto*), an introduced species, across North America (Sullivan et al. 2009). In addition, volunteers are identifying patterns in massive data resources. For example, a Dutch school teacher discovered a new type of nebula while participating in the Galaxy Zoo (www.zooniverse.org).

The ability to recruit, engage, and reward individual citizens and encourage them to participate in global-scale citizen science networks has been dramatically improved by advances in four related information technologies (Friedman 2005): (1) computers are everywhere; (2) the Internet and Web browsers employ global standards for passing information among computers; (3) the pervasive installation of fiber-optic cabling has globalized computer networks; and (4) development in information description languages, data-management processes, and software application integration together have created seamless workflows for contributing, accessing, and processing almost limitless data resources. These advances have fundamentally transformed how we obtain, process, and communicate information in almost all aspects of our daily lives. In this chapter I present the framework to apply this global Internet infrastructure to advance citizen science engagement opportunities that provide the framework for research, education, and dissemination of information at global scales.

Bioinformatics uses tools developed in the computer sciences to collect, organize, and analyze biological data. Bioinformatics has made major advances in genetic research and medicine and is used across all of the sciences—from astronomy to particle physics and from the environment to human social networks. By taking advantage of the increasing access to massive quantities of well-managed data, computationally intensive techniques such as interactive Web-based exploration and visualization tools and machine learning techniques can be used to analyze multivaried systems.

A major feature of bioinformatics is the development of a data-management cyberinfrastructure that includes all of the data maintenance processes and procedures related to data administration, quality control, security, and storage, along with the variety of procedures that "deliver" or provide access to the data (Martin et al. 2009). This chapter describes the interrelated data-management processes that all citizen science projects should implement if they wish to take full advantage of existing bioinformatics and cyberinfrastructure resources. I address strategies for managing and archiving data as well as the Cornell Lab of Ornithology's techniques for making data discoverable via metadata, allowing data to be accessible in an interoperable format, i.e., the ability to interpret data that crosses system or domain boundaries. I also provide examples of how the tools we create allow contributors to explore, synthesize, and visualize citizen science data and permit professional scientists to perform complex analyses on the same data sets.

Data Management

Traditionally, scientific or engineering data-management practices have remained local, idiosyncratic, and oriented to current use rather than to preservation, curation, future access, and reuse (Borgman 2007). The fading memory of completed data sets leads to data entropy, that is, the information content associated with the data is lost over time (Michener 2006). Adapting a sound data-management strategy reduces the risk of data loss and allows for efficient use and reuse of data. Good data-management practices also lead to rich opportunities for collaboration and for synthesis of data across projects or domains. Additionally, because many funding agencies now require plans for data management, solid strategies improve the chance of success in obtaining grants for continuing research. Data management is particularly important for citizen science activities because information collected by participants is essentially a public good, and participants expect their data to be housed in permanent collections. Here I identify several important components of an effective system for managing citizen science data.

Databases

Data in citizen science projects should be stored in database management systems (DBMS). A DBMS is an organizational structure that stores, processes, and manages large amounts of data in a systematic way and that has the potential to provide multiple users with access to the same information. A DBMS is distinguished from a spreadsheet, which is designed for numerical analysis and manipulation. When researchers attempt to use spreadsheets for long-term data management they must recopy data over separate data files and typically have difficulties detecting data errors or maintaining original data. Database design and the adoption of sound DBMS practices should occur during the planning of a citizen science project, because the DBMS will provide the foundation for all key applications including data gathering, data quality review, data visualization, and analysis.

To test a project's functionality before releasing a fully developed cyberinfrastructure to a broad audience of participants, "pilot" projects should be implemented. Design shortcomings that are uncovered and fixed during a pilot save time, cost, and aggravation for all who work with the database including programmers, database administrators, and data analysts.

Well-designed relational databases are scalable and can easily handle data collected from single or multiple projects. Citizen science data can be efficiently stored in relational databases consisting of separate tables that have explicitly defined relationships with each other and whose elements may be selectively combined. These separate tables reduce redundant information *within* a project by decreasing the number of repeated identifiers (e.g., participant identifiers). They also decrease redundant information *across* projects by allowing direct linkages to data without the need for copying data, which increases the chance for creating inconsistent data (Borer et al. 2009).

Tables in a relational database should be designed to hold records that are the fundamental elements of the project and that represent entities found in the real world (e.g., users, locations, observations, data-collection trips). The declared relationships between tables should match the natural relationships between the entities they represent. For example, because a data-collection event happens at a specific location, a collection event record should reference a location record. Furthermore, because a data-collection event is recorded by a primary user, each collection event record should reference a record in the user table. Finally, because every observation is part of a data-collection event, each observation record should reference a record in the collection event table. Once this combination of tables and their relationships are defined correctly, they will follow a logic that can be understood and followed by those who work with them. For example, programmers and analysts will find working with the database to be intuitive, and their data queries will be easily posed because the database is a "natural fit" with the entities it represents.

Citizen scientists collect observations using a myriad of mechanisms as diverse as measuring rainfall or water quality, recording birds visiting a backyard feeder, or discovering new comets or entire galaxies. All these observations share common thematic components that describe some entity along with traits or processes involving that entity. For example, many citizen science projects identify the presence of a species (the entity), its age (trait), or the behavior (process) that was observed (Kelling 2008). Experience has led to development of a "single observation" relational database format for the majority of the Cornell Lab of Ornithology's citizen science projects. This enterprise database uses a hierarchical model that is based primarily on four relational data tables (Figure 3.1). The user table contains a list of project participants with data fields such as name and contact information. The location table provides detailed information about where the observations were made. The collection event table contains (1) information on the protocol and effort made to collect the data, (2) information

Relational data model

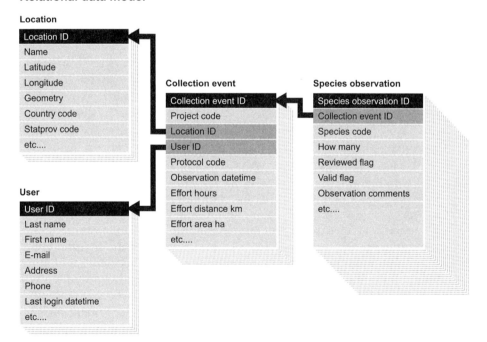

Figure 3.1. The Cornell Lab of Ornithology data model for citizen science. All data are stored within a relational database that uses four relational data tables. The user table contains information about project participants. The location table provides information about where the observations were made. The collection event table contains information about how the data were collected. The observation table includes the identification of the species and what the observations were. The information within the tables is organized logically, working with the database is intuitive, and queries can be easily developed.

about which participants from the user table made the observation, and (3) information about which location from the location table was the site of the observations. The species observation table includes information about the taxonomic identification of the species, the number of individuals observed (count data), information on a specific individual (e.g., weight, size) if those data were collected, and data-quality flags that indicate whether an observation was reviewed (either by a filter or an expert). Each record in these four tables includes a unique identifier that allows it to be linked (i.e., related) to records in other tables. Finally, whereas primary tables are able to organize the majority of data being collected, auxiliary tables have been implemented as a way to allow individual projects to effectively widen the base tables with extra data fields customized to the needs of an individual project.

To date (February 2011), the Cornell Lab of Ornithology maintains a database of bird observations from 281 unique projects with data gathered by 121,875 individuals from 938,078 unique locations. Together, these databases comprise 9,158,013 individual data collection events and 101,718,295 species observations.

Data Archives

The goal of archiving scientific data is to locate data together with the methods used to gather the data in a long-term storage device. This is accomplished by moving all data to a secondary storage tier (with the emphasis on data retention, discovery, and reuse.) Archiving is more than simply making occasional data backups; rather, it provides a complete description of the data via metadata (see below) along with easy access to the data for reuse and synthesis. Unfortunately, although archiving is one of the most critical aspects of a sound data-management strategy, it typically receives little attention in citizen science or other disciplines.

A proper archive for a citizen science project should clearly identify what data exist and how to access them and provide a clear description of the information contained in the data. Specifically, the value of the data archive is not only what data are stored but how the data are stored. Therefore, the type of storage device, data format, and type of access to the stored data influence the ultimate value of the data (Brunt 2000). Ultimately, the archive should become the reference version of the data and thus the basis for all subsequent work. For this to occur the archive must provide sufficient documentation to permit effective use by anyone unfamiliar with the original project or the particular protocols used during the project (Munson et al. 2009).

In addition to making project archives available as "working copies" of project databases, the Cornell Lab of Ornithology has adopted an archival strategy to ensure that no data are lost. First, every month each database is backed up in its entirety, initially to locally running disks and then off-site to a tape/disk enterprise backup system. Any and all changes made to the database are also simultaneously written locally to "change logs" on two different disk volumes. The logs are copied to the off-site system every four hours, and because they grow quickly (currently more than 13 GB/day), once or twice each week they are packaged into "incremental backups" and copied off-site. Therefore, even if the database machine suffered a complete loss, only the most recent four hours of user data activity would be irretrievable. Finally, we perform database restoration exercises several times each year to simulate total loss of the database machine. These exercises prove the viability of the backups, improve our restore procedures, and prepare us for an actual restore. In the fourteen years that our system has been in place no data have been lost or destroyed, even though we have experienced several hardware failures and many data migrations to new versions of database software.

Metadata

Metadata are facts about data that are used to document an informational resource. Just as library catalogs detail information about a book and how to locate that book, metadata for citizen science data sets describe the collections and how they can be accessed. Metadata provide information on the origin, ownership, quality, context, data structure, and accessibility of projects, using a common terminology and set of definitions that prevent loss of the original meaning and value of the resource (Michener 2006). This common terminology is essential because citizen science projects collect disparate information, occur at a variety of scales, and are dispersed globally. Thus, without metadata, discovering that a resource exists, what data were collected, and how to access and properly use the data would be impossible.

Metadata provide the following:

- A standardized format that describes the projects and protocols that gather observational data
- An accurate description of the actual data
- A description of spatial attributes, which should include bounding coordinates for the specific project, how spatial data were gathered, limits of coverage, and how the data are stored
- A complete description of the taxonomic system used by the project, with references to methods employed for organism identification
- A description of the data structure, with details on how to access the data or how to access tools that can manipulate the data or both (i.e., visualizations, statistical processes, and modeling)

A variety of metadata standards are used to organize information. For the purposes of citizen science, two interchangeable metadata standards are recommended. The first is the Biological Data Profile, developed by the Biological Data Working Group of the Federal Geographic Data Committee

(FGDC) to increase compatibility in the development, use, sharing, and dissemination of biological data. The specific goal is to define all information required by a user to determine what variables are stored in the data set, the data quality, and how to access the data. It explicitly describes data sets through a series of interrelated metadata elements that provide a detailed description of the data set's contents, and it yields a detailed data-set profile.

The Ecological Markup Language (EML) was developed by the ecological community to provide a common structure that allows ecologists to discover ecological data and to provide sufficient information for a researcher to use the data in a scientifically correct manner (knb.ecoinformatics.org/software/eml/). EML metadata are based on the FGDC standard, and descriptors are organized into classes that describe the data set and its research origin, data structure, status, and accessibility. The metadata categories in EML are extensive, and metadata descriptions can be constructed on either the entire data set or a subset (attribute) of the data. EML is highly structured and provides the basis for organizing and tagging elements through eXtensible Markup Language (XML), which allows the data to be easily transmitted between applications and computer platforms.

Ensuring Data Quality

Murphy's Law, which states that anything that can go wrong will go wrong, can easily be applied to citizen science. For example, as the number of contributors to a project increases, the opportunity to have similar information described differently also increases (e.g., "December 3" vs. "3 Dec."). Data-entry forms that use a controlled vocabulary—for example, a standardized terminology for dates, taxonomy, or behavior classes—eliminate much of the variability in how people report their observations. Such forms ensure that people report their observations in the same language and enhance the ability to organize, store, retrieve, and analyze those data. The benefit of standardization usually outweighs the cost of restriction.

Accuracy of identification also plays a major role in data quality, and citizen science participants exhibit significant variability in this area. While a significant multibillion-dollar industry is training individuals to become better bird observers, this is not the case for most taxa. Consequently, citizen science projects must develop significant checks and balances to ensure that species identifications and related information contributed by the public are correct. For example, spatial, temporal, and quantity filters constructed within the projects' application architecture can provide a data-verification system that instantaneously evaluates submissions before they appear in the database. These filters can function behind the scenes and can serve multiple functions. First, they provide the user with a list of organisms, behaviors, or other measurable characteristics that are most likely to be expected on the reporting date at a specific location. Next, filters can flag any unusual records in the database, such as atypically high numbers or organisms outside of their normal range or season. Finally, records that have been "flagged" can be sent to an expert for review and possible acceptance. When the expert accepts a "flagged" record, the "flag" is removed from the record in the database. If the expert rejects the record, then the "flag" remains on the record, but the record remains in the database.

Data Interoperability

Observations made by networks of citizen scientists can be used most effectively if they are combined with other types of data (e.g., land cover, human demographics, climate data) that potentially impact how, when, or where the observations were made. For example, combining observational data on bird occurrence with data on habitat type allows researchers to find correlations that allow more accurate predictions of the occurrences of organisms across broad spatial and temporal scales. But currently these different data resources are difficult to locate, stored in heterogeneous data structures, and obtained through different access pathways. Combining data requires adoption of formal metadata and semantic frameworks to allow a researcher, as opposed to a programmer or informatics specialist, to intuitively synthesize disparate data resources via a straightforward user interface (Jones et al. 2006).

The Cornell Lab of Ornithology, in partnership with thirty-seven other governmental and nongovernmental organizations, has created the Avian Knowledge Network (AKN) (www.avianknowledge.net) to organize and make available a massive data resource on bird occurrence. Each data set within the AKN has complete FGDC biological data-profile metadata providing sufficient descriptive information to document its structure, contents, and use constraints.

The AKN uses a data warehouse, where data have been converted into a standard data model. Creating a data warehouse requires considerable effort up front, but it eases subsequent use of data by a wide range of analysts, including academic researchers, conservation organizations, and citizen science participants. The model employed by the AKN is similar to other efforts to integrate diverse data from multiple fields of ecological research (McGuire et al. 2008). Our warehouse is multidimensional and consists of an event table (i.e., the information detailing the observation of a bird) and multiple predictor tables (i.e., variables that impact the observation of the bird) (Kelling et al. 2009).

Bringing together bird observational data into the observation event table was challenging because data collection techniques are diverse, the original data resources were widely dispersed and owned by a variety of organizations, and the data formats varied dramatically. To overcome these issues we developed a common data schema, the Bird Monitoring Data Exchange (BMDE) (Lepage et al. 2005), which captures as many data elements as possible to describe the bird observation event (i.e., who made an observation, what was observed, and how and when an observation was made). The BMDE schema is based on Darwin Core, which provides a suite of descriptors to identify the commonality in the content of biological collections and observation data (Wieczorek 2007), with extensions to describe characteristics of bird occurrence. In the BMDE schema a record in each data set corresponds to a checklist that an observer used to mark the number of birds of each species detected. One checklist is submitted for each sampling event. Each sampling event comprises a series of observations, which record both the presence and absence of birds. Observations of bird presence can consist of multiple occurrences (i.e., counts) of that species made during the sampling event. Most data sets within the AKN assume that the observer has reported all birds detected during a sampling event, except when specified as a different type of count. Consequently, when a species was not reported, observations of the bird's absence can be inferred.

While bird observation data come from multiple sources, each environmental attribute (i.e., predictor variable) associated with a bird observation came from a single, uniformly collected data set. Collection mechanisms for these variables ranged from remote sensing (e.g., land cover) to anthropogenic surveys gathering human demographic information. Because all bird observations and attributes have latitude, longitude, and date as shared context, we were able to join observations of species with observations of environmental features at individual locations.

The AKN has begun to make reference data sets available that include observations of birds linked to many predictor variables. For example, the largest data set housed at the AKN is the eBird Reference Dataset (Munson et al. 2009), and in June 2011 it contained more than 41 million observations gathered during almost 2.5 million sampling events at more than 320,000 locations. All checklists submitted from the contiguous forty-eight United States are included along with detailed location-based covariate information. Additionally, all other eBird checklists submitted from North, Central, and South America, including the Caribbean islands, are available, although they do not contain location-based covariates. The eBird reference data set contains information regarding the sampling event (i.e., when, where, and how the observations were made), environmental covariates that are felt to be generally useful for modeling species occurrence (i.e., elevation, land cover, climate, housing density, and human population), and a series of spatial pattern statistics (i.e., habitat patch sizes, distance between habitat patches, and patch densities within predefined regions around the observation). The eBird Reference Dataset provides a powerful resource for understanding the determinants of species distributions and why distributions change through time, information critical for conservation and management.

Visualization and Analyses

Based on the increasing availability of massive volumes of well-organized data, the sciences are becoming increasingly data driven and computationally intensive. This new data-analysis paradigm in scientific research is called data-intensive science (Hochachka et al. 2007; Hey et al. 2009). For biodiversity research, data-intensive science addresses the challenges of manipulating enormous volumes of data that are analyzed in increasingly complex computing environments (Kelling et al. 2009). Data-intensive science functions through scientific workflows, which are processes used for accomplishing a scientific objective and expressed as a series of tasks and dependencies (Ludescher et al. 2009). A data-intensive scientific workflow (Figure 3.2) must include the following steps. First, data are collected from various sources and validated for quality. Next, disparate data are synthesized and iterative exploratory techniques are used to discover interesting patterns in the data. Additional information can often be harvested using model-based explorations. By interpreting and analyzing these models, truly novel and surprising patterns can be discovered. These patterns in turn provide valuable insight for concrete hypotheses about the underlying processes that created the observed data.

Data-intensive science processes can take advantage of the tremendous volume of data being gathered through vast networks of citizen science contributors. For example, new computationally intensive modeling techniques are being developed that provide accurate representations of species occurrence across broad spatial and temporal expanses (Caruana et al. 2006;

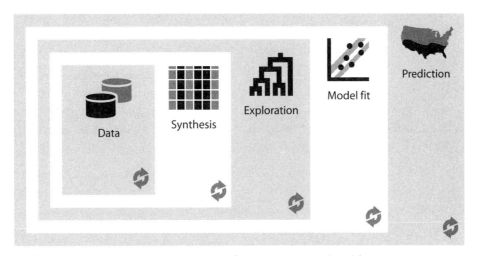

Figure 3.2. An iterative, data-intensive workflow. (1) Data are gathered from various sources and (2) synthesized; (3) multiple exploratory analyses characterize the data and suggest approaches to modeling; (4) model fitting leads to (5) predictions and hypothesis testing. At any step it may be necessary to backtrack, taking some other route forward.

Fink et al. 2010; Sorokina et al. 2009; Sullivan et al. 2009). In addition, with growing efforts to reform science education that call for engaging students in data-driven inquiry-based problem solving (Brew 2003), access to citizen science data through data-intensive science workflows can make it feasible for students to conduct meaningful scientific explorations through intuitive and interactive visualization tools that do not require understanding the complexity of the underlying data structure and analysis processes. Finally, to provide decision makers with the tools they need to explore data and assumptions through a series of "what if" scenarios, data-intensive science will allow them to evaluate complex multivariate systems that are described by numerous parameters and then to make accurate decisions.

❦

Organizing and providing resources to gather, manage, discover, and access large quantities of observational data provides the foundation for new discoveries and a better understanding of natural systems. New bioinformatics initiatives to organize observations gathered through citizen science projects provide the opportunity for the creation, implementation, and sustained management of an integrated and comprehensive data-curation strategy. This task is not easy and the challenges are often underestimated, requiring the proper coordination and cooperation of professionals from widely diverse disciplines including scientists, informatics and computational specialists, application developers, database managers, and visualization specialists. The creation of this new data-management framework will, however, overcome the data dependencies that disciplines or individuals within a discipline often embrace, because it provides an obvious benefit of data access, security, and synthesis needed for a broad range of users that includes scientists, educators, students, land managers, and the interested public.

Acknowledgments
This work was funded by the Leon Levy Foundation and the National Science Foundation (ITR-0427914, DBI-0542868, DUE-0734857, IIS-0748626, and IIS-0612031). Thanks to Will Morris for developing the figures.

4

Growing the Base for Citizen Science

Recruiting and Engaging Participants

MIYOKO CHU, PATRICIA LEONARD, AND FLISA STEVENSON

The birth of a citizen science project may begin with scientific questions and educational objectives, but its ultimate success hinges on a practical question: Given the scientific requirements of large geographic coverage and repeated observations at the same site, will enough people participate to reach the project goals? Unfortunately, it's not true that if you simply build it, they will come. First, prospective participants have to find out about the project. Then they must decide whether to participate. Next, they have to learn how to participate. Finally, they must volunteer their time to collect and submit data. Recruiting hundreds of thousands of people to participate in national citizen science projects is a challenge. In this chapter we explain strategies for recruitment and retention, such as providing an experience that is rewarding to participants, crafting messages that are relevant to targeted groups, spreading word of the opportunity through diverse channels, and creating community. Ideally, these factors should be considered during project development and continue to evolve for the lifetime of the project.

Below, in the section on "Creating Projects for People," we explain the importance of understanding participant motivations and designing projects to appeal to targeted groups. A case study explains how the eBird project offers a rewarding experience for participants by providing features that dovetail with user needs and interests.

In "Spreading the Word," we describe ways to publicize citizen science projects through national and local media campaigns, Web communications, and engagement of volunteer "ambassadors." A case study explains how the Great Backyard Bird Count mobilizes tens of thousands of people in the United States and Canada for an annual four-day count.

In "Building Community," we explore how participants and program staff broaden and deepen their relationships with one another through on-line communities and on-the-ground community involvement. A case study focuses on the success of the Celebrate Urban Birds project in partner-ing with community organizations to reach and engage diverse groups of people. At the end of the chapter we consider current and future challenges for recruiting and engaging participants.

Creating Projects for People: Participant-Centered Recruitment

When the Cornell Lab of Ornithology first began recruiting citizen science participants, typical promotional messages asked people to participate in order to help others (e.g., to help scientists by collecting data and help the birds by contributing to conservation). This was consistent with traditional volunteer recruitment approaches that appeal to people's altruism in volun-teering their time for a good cause (Callow 2004).

Research findings indicate that although volunteers are motivated by a desire to help others, they also perceive benefits to themselves, such as the satisfaction of being productive or the enjoyment of interacting with others during volunteer activities (Wymer 2003). Although altruism draws people into volunteer projects, a match between volunteers' motivations and the activities they are asked to perform better predicts volunteer sat-isfaction and intentions to stick with a project (Stukas et al. 2009). By ex-panding messages to include the quality of volunteer experience, nonprofit organizations can appeal to a wider target audience than with traditional messages alone (Callow 2004). This involves considering both motivations and perceived benefits, and applying principles that treat volunteers as cus-tomers, with the goal to better serve these customers than competitors do (Wymer 2003).

For example, the NestWatch project needs volunteers who can monitor bird nests safely and accurately. When the project launched, it targeted participants in The Birdhouse Network (the project's predecessor) as well as individuals who were already monitoring nest-box trails—people who knew the process, had a personal commitment to it, and were producing lots of data. In addition to emphasizing the value of contributing data, the project created tools of value to nest-box monitors such as the ability to record, organize, and view their own data online and to explore data sub-mitted by other participants.

Communications strategies that consider a greater variety of motiva-tions and benefits can help reach a larger potential pool of people when projects require broad participation. For example, Callow (2004) found that older people are often treated as a homogeneous demographic group in volunteer recruitment publicity, and opportunities are therefore missed to appeal to people with different personalities (e.g., people oriented vs. task oriented) and desires (e.g., get to know other people through volunteering vs. learn new skills). A more effective approach can be to segment the targeted audience, launching multiple campaigns with messages tailored to be relevant to each segment. For example, the Celebrate Urban Birds project launched three campaigns to engage people with different interests: one focused on promoting participation through the arts, another on gardening for birds, and another on contributing to science (see Chapter 13).

Participant-centered approaches enhance recruitment by appealing directly to participants' interests and motivations. Providing multiple ways to engage in the project opens doors to people with diverse interests and backgrounds. It leads to greater participant satisfaction because people can choose the kind of experience they enjoy most. And it provides insights about audience niches for future growth and improvement. Research has shown that promotional messages are most effective and satisfaction is highest when messages and benefits of volunteering match the volunteers' motivations (Clary et al. 1998).

Although recruitment is crucial for growing the base of participants, retention is equally important, especially because long-term data from specific participants or locales have the greatest scientific value. Therefore, acquisition of new participants is just the beginning of a cycle that involves creating materials or messages to welcome participants to the project and then continuing to engage them over time—through e-newsletters, print publications that provide support and share news and results of the data they have contributed, or interactive websites that provide ongoing support and project results. The cycle leads to new growth when participants spread the word among friends and acquaintances, or when cross-promotional efforts lead participants to join additional projects.

Case Study: eBird Soars with Tools Tailored for Birders

How do you convince bird-watchers to voluntarily report millions of bird observations across an entire continent in a consistent way, year-round? That was the challenge when the Cornell Lab of Ornithology and the National Audubon Society launched eBird (ebird.org) in 2002.

The program began by posing key scientific questions that could be answered if enough birders participated and by creating the scientific protocols to ensure valid and useful data collection. But convincing thousands of birders to submit their observations—the lifeblood of eBird—was an evolving process. From 2002 to 2005, participation in eBird held steady at around 200,000 checklists per month. By 2010 the project was receiving more than 1.3 million reports each month, more than doubling its earlier monthly rate during peak months (Figure 4.1). The growth began with a concerted effort to build eBird into the kind of tool that *users* wanted—while still collecting the information desired by scientists.

Indeed, the 2005 launch of "eBird 2.0" included a new generation of features specifically designed for users. To create this new version, the Cornell Lab of Ornithology hired project leaders Brian Sullivan and Chris Wood, expert birders who are deeply plugged into the birding community. They gathered input about desired features by e-mail, in personal conversations, and at venues such as birding festivals and conferences.

The eBird team began by adding online tools that allowed users to keep track of life lists and state lists, to bulk upload data that had been recorded in other formats, to report and display information about rare birds, and to create output charts and graphs for their own sightings as well as for regional and national trends.

The team also built partnerships with state and regional conservation organizations and, to date, has created more than two dozen customized eBird portals. For example, a Canadian portal has proved critical to successful coverage of North America because national identity is a major factor in participants' desire to join. Specialized portals are also important; for example, Chicago's Bird Conservation Network eBird portal collects data to answer questions about declining grassland birds. Likewise, both New York and Florida use eBird for regional projects, pinpointing areas used by endangered Red Knots and using these data to inform decisions about local development. Data that participants enter through this portal

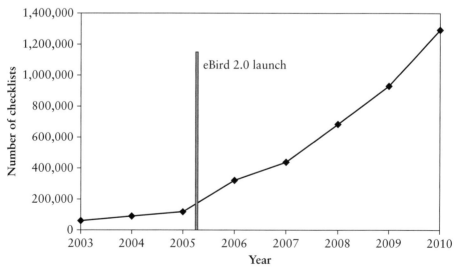

Figure 4.1. Increase in the number of checklists (counts during a bird-watching event) submitted to eBird since the project began in 2003.

also go into the main eBird database for everyone to use. As a result of the 2005 initiatives, eBird participation followed a steady upward path as more groups and individuals made eBird their own.

Along with development of online tools and customized portals, project leaders became traveling eBird salesmen. "The thing that works better than anything else," says coleader Chris Wood, "is meeting with people in person and showing them what eBird can do for them. Once we establish a relationship with a few people in a state, they get so interested they spread the word among bird clubs and other groups in the state or region."

Though face time with potential users is vital, other communication avenues are also part of the mix. Nurturing the relationship with existing eBird users includes electronic newsletters, website articles, recognition for high-volume contributors, an eBird blog, Twitter feeds, a Flickr pool for rare bird photos, and an RSS feature to alert participants to something new on the site. Keeping users happy is important because of the value of long-term data from the same localities, and because personal communication continues to be one of the primary ways that news is spread. "Everybody who uses eBird data is another promotion for eBird," says Wood. "That helps us get more people to use it." Although we have not formally evaluated which of these tactics is most successful, the cumulative response has been remarkable.

Spreading the Word: The Power of Messages and Multifaceted Approaches to Publicity

Publicity campaigns for citizen science projects have changed dramatically in the past decade, with multiple strategies for reaching different audiences in increasingly diverse ways. These campaigns are also capitalizing on new Web technologies, keeping pace with major shifts in how people receive information.

For example, when the Great Backyard Bird Count citizen science project was launched in 1998, the Cornell Lab of Ornithology and National Audubon Society generated publicity primarily by faxing press releases to newspapers one at a time. Today, regionally tailored press releases can be nearly instantaneously e-mailed to thousands of newspapers, magazines, and radio and television stations. The releases also provide links to the Great Backyard Bird Count website and its press room, from which media outlets can obtain high-resolution photos and sounds, and contact any of the GBBC hosting organizations to connect with local participants for stories about the event. Participants also help spread the word on the Internet through blogs, Facebook, and Twitter—and even by posting bird photos and videos on Flickr and YouTube. Flyers go out electronically for libraries, clubs, nature centers, and stores to post in their communities. Volunteer "ambassadors" mobilize their own communities by downloading Power-Point programs and giving presentations, or by printing materials from the Web to host local events to tally birds and submit data.

Effective messages often pair a call to action with an enriching or enjoyable experience for the participant. For example, in 2010, a press release for NestWatch emphasized that collecting information from participants across the continent over long periods of time allowed scientists to detect widespread changes in bird breeding biology, some of which may be related to climate change. A quote from the project leader captured the scientific significance of the findings: "Data show some species, like the Tree Swallow, are laying their eggs more than a week earlier than they did just a few decades ago. That could spell big trouble if hatch dates get out of sync with the availability of food." But the press release also alluded to the project's "fun" factor, highlighting a New York kindergarten class that collected information about bluebirds nesting on school grounds. It emphasized the personal value of participation by quoting the teacher: "What a marvelous experience for all of us to enjoy and learn!"

In some cases, the call to action isn't even related to science but asks volunteers to create and submit bird art, anecdotes, or photographs for contests, or to share experiences such as birding with a grandchild. Messages can be tailored to suit the culture and demographic. For example, Celebrate Urban Birds often mentions health and wellness benefits of bird-watching as a counterbalance to urban stress: "Take a break with the birds."

Because science can be intimidating to some audiences, messages often emphasize the ease of joining or mention that "anyone" can participate. Another effective approach is to use narratives, which can help engage non-expert audiences through storytelling (Caldiero 2007). For example, Celebrate Urban Birds launched a contest inviting people to send in their photos and stories of "Funky Nests," bird nests found in bizarre places such as on clothespins, satellite dishes, barbecue grills, and stoplights, or in old boots. Participants submitted more than 500 photos, stories, videos, and drawings, most of which were posted on the Celebrate Urban Birds website. In addition to showcasing participants' photos, the website provides information on how to watch bird nests and participate in other activities that celebrate urban birds.

Case Study: Publicity Fuels the Great Backyard Bird Count

In just 4 days, more than 65,000 people across North America counted birds for the 2011 Great Backyard Bird Count (GBBC). They watched backyard feeders. They took guided hikes on nature trails. They counted birds in schoolyards, parks, and neighborhoods. They submitted 92,200 checklists, tallying 11.5 million individual bird observations of 596 species.

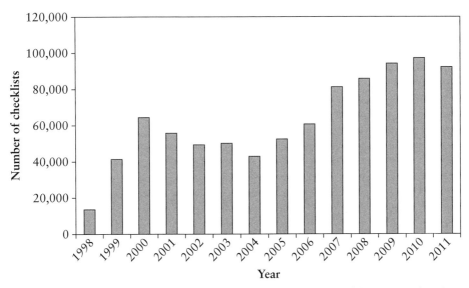

Figure 4.2. Number of bird checklists submitted for the first 14 years of the Great Backyard Bird Count.

The GBBC has come a long way since its first outing in 1998. It was developed as part of a new bird observation database, BirdSource, created by the Cornell Lab of Ornithology and National Audubon Society. BirdSource was an early attempt to tap the Internet boom by collecting reports from birders across the continent. Although the name *BirdSource* is no longer active, the pioneering work of this endeavor has fueled all of the Cornell Lab's citizen science work since.

"With the GBBC, we were trying to test volume," says Steve Kelling, director of information science at the Cornell Lab. "How much could our databases handle? How many checklists could we process? Could we use the data to build maps instantly?" The first GBBC took in 11,700 checklists and was considered a great success. Although a few online bird surveys had been done before, this was easily the largest of its time. But in order to fully answer questions about how much data could be processed online and to give patterns in the data a chance to stand out, the GBBC needed to grow. Participation peaked after three years and then started to decline (Figure 4.2). Spreading the word about the GBBC became an increasing communications challenge.

Media Messages
Key media messages stress that the GBBC is free, easy, fun, family friendly, and not time-consuming. Promotional messages also underscore the value of the project to science and conservation, emphasizing that it serves as a

sentinel project by producing observations of changes in bird distribution and abundance that targeted research could follow up on. Survey results show that these messages are consistent with the impressions and motivations of participants. In response to open-ended questions, many participants said they found the GBBC website to be user-friendly and the project easy to do. In addition, 98% said that they would participate again, showing that once people find out about the GBBC, they come back. Fifty-six percent of respondents said that one of the main reasons they took part was that they simply enjoyed watching birds; 45% said that they wanted to contribute to science and bird conservation. As one North Carolina participant put it: "I was thrilled to be part of something that would help shed light on the environment and the impact that humans have on this earth. It was a chance to use my hobby for a greater good."

Media Strategies: The Local Angle

To interest local media in a national event such as the Great Backyard Bird Count, we create media materials that show how the story is relevant to local backyards, neighborhoods, parks, and nature centers. States, provinces, and towns that submit the most checklists are recognized online to generate local interest and pride. Regionally tailored news releases highlight past participation from the area and any significant trends to watch for, such as an influx of northern finches or expansion of an invasive species from the south.

Local people also make the GBBC a local story. The project recruits "media ambassadors," participants who are willing to be interviewed about their experience, and also connects interviewees with reporters. Media stories often feature local participants or focus on local GBBC events and results. Some news stories have also focused on how citizen science data on bird distributions may shed light on climate change—a theme of GBBC news releases in both 2009 and 2010.

More than 1400 GBBC articles appeared in mainstream media in 2010. Hundreds of local newspapers, magazines, radio, and TV stations have featured the event. Highlights of media placements in recent years include CNN, *The Martha Stewart Show*, National Public Radio, and the game show *Jeopardy!*

Talking to Communities

Aside from media coverage, successful publicity also relies on keeping past and present participants well-informed and creating a shared sense of community. The GBBC staff issues a series of colorful electronic newsletters before, during, and after the count, reminding participants of the dates, highlighting new project features, announcing winners of photo contests and prize drawings, and asking recipients to let others know about the event. Publicity efforts also target bird clubs, nature centers, garden clubs,

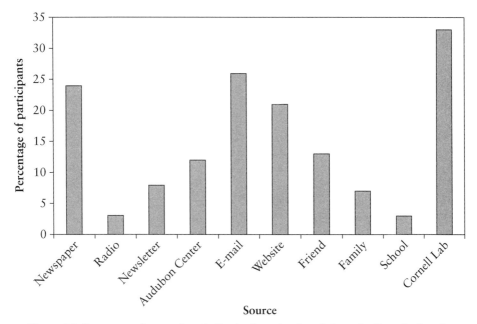

Figure 4.3. Percentage of respondents indicating how they heard about the Great Backyard Bird Count, based on a 2009 participant survey. Multiple responses were permitted.

bird bloggers, libraries, youth groups, homeschoolers, and others. A concerted effort to expand such distribution lists began in 2005 and GBBC participation has grown every year since. Surveys indicate that people hear about the event in many ways—through electronic and print media as well as word of mouth (Figure 4.3).

Ambassadors

Hundreds of volunteer GBBC ambassadors distribute posters, write articles, hold birding workshops, and talk up the GBBC on personal blogs and websites. Parks and nature centers host GBBC seminars and bird walks before and during the event. Classrooms get involved; for example, grammar school students in Fultondale, Alabama, shot GBBC "commercials" and posted them on their classroom blog.

Wild Birds Unlimited stores sponsor the GBBC with franchise owners holding birding workshops, leading bird walks, and doing interviews with local media. Store owners are urged to use the GBBC to strengthen community ties, involve the community in creating a snapshot of the continent's bird populations, and attract more customers.

Website Features

The GBBC website is a vital communications tool. It contains instructions, bird identification tips, a kids' page, a photo gallery, a list of local events, a downloadable participation certificate, and prize drawings. Participants submitted nearly 6000 images for the photo contest in 2011. All these communications strategies are intended to produce a memorable birding experience, the success of which was captured by this survey comment: "I enjoyed the feeling of being part of a coast-to-coast birding community!"

Building Community: How Partnerships and People Give Projects a Life of Their Own

Citizen science projects are inherently about partnerships—collaborations between scientists and volunteers. But successful projects also become much more than that. They evolve into a community, a family of people who participate in a common cause together. Participants connect with one another, sharing their data not only through the project's visualization tools but also through stories, photos, and insights about their experiences. They become the project's best advocates, volunteering not only to watch birds for the project but to post flyers, send press releases to their local media, and teach others how to participate.

Project staff help facilitate these communities in numerous ways. Project FeederWatch publishes stories about participants in newsletters and on the Web and produces an annual calendar featuring photos and quotes from participants. Some projects host online forums enabling people to ask one another questions and share advice. And as we described above, the Great Backyard Bird Count recruits hundreds of volunteers to serve as "ambassadors" who serve as a voice for the GBBC in their community.

Partnerships with other organizations can quickly bring projects directly to communities through existing groups, such as after-school programs, senior centers, and nature centers. Partnerships help connect people to the project in a familiar social context aligning with interests that are important to them. The Celebrate Urban Birds project has shown how partnerships can increase not only the numbers of people a project can reach but the depth of the experience as well.

Case Study: Communities Adopt and Adapt Celebrate Urban Birds

To reach urban participants, the Celebrate Urban Birds project, in addition to attracting individual participants, engages community organizations as partners. The goals of the project are to increase audience diversity and to collect data on sixteen species of urban birds to understand how these species use different habitats across the continent. During 2007–2009, more than 5000 partner organizations signed up, helping Celebrate Urban Birds to distribute more than 110,000 participant kits. More than 80% of the partner organizations are working with underserved audiences. Partnerships have helped the project reach niche communities such as corporate wellness programs, scouts, 4-H groups, camps, schools, after-school programs, organizations working with autistic children, battered women's shelters, faith-based groups, senior centers, rehab programs, and others.

Celebrate Urban Birds also developed a mini-grant program that enabled organizations to envision and offer meaningful activities in their communities. In 2009, the project invited proposals for twenty-three grants of between $250 and $500 each. Publicity around the availability of the mini-grants raised awareness of the Celebrate Urban Birds project and inspired community groups to consider the possibilities. The project received more than 650 applications. The Cornell Lab of Ornithology invited its members and eNews subscribers to help select the winners by voting online for a favorite project. This raised awareness of the project's impact and generated online donations that helped fund additional proposals.

Winners included the Alzheimer's Alliance of Smith County, Texas, which used the funding to bring participants with Alzheimer's disease and fifteen elementary school students together for an event called "Make Memories Fly." They made birdhouses together, set up bird feeders, planted sunflowers, and watched birds—activities that taught the children about birds and human diversity, while helping Alzheimer's participants evoke past memories of camping, hiking, and bird-watching. The Martin Luther King Community Center in Indiana incorporated activities celebrating urban birds in their Second Time Around program for grandparents who are raising grandchildren. The project helped children connect with nature and gain confidence through gardening activities that beautified the community center. The Queen City Creamery in Maryland promoted science, art, and outdoor activities celebrating birds and awarded free ice cream cones to anyone who submitted a data form for the project.

The Celebrate Urban Birds staff provided mini-grant winners with a sample news release, e-mail message, and flyer that they could tailor for their local media and social networks. Newspapers featured stories of winners in their community, giving recognition to the organization's efforts and raising awareness of urban birds. The Celebrate Urban Birds website also showcased the work of partner organizations. Through the mini-grant program, Celebrate Urban Birds reached underserved groups and inspired other organizations like them to explore the natural world.

"Celebrate Urban Birds is environmental education and environment-based education that is multi-generational and inclusive," one participant wrote in a survey. "We've seen that this citizen-science project emphasizes a sense of place, how we fit into this place and system, and a methodology for saving our place. There is a cool stewardship dynamic that comes easily."

Current and Future Challenges

During the past decade, communications professionals at the Cornell Lab of Ornithology have worked closely with project leaders and principal investigators to develop protocols and approaches that work for scientists as well as participants. Communications strategies for citizen science projects have evolved from traditional messages focused on volunteer altruism to include participant-centered approaches with multiple campaigns and experiential enhancements built into the projects. Growth has increased rapidly for projects such as eBird, the Great Backyard Bird Count, and Celebrate Urban Birds, which now reach more than 200,000 people in total.

When we consider, however, that some 41.8 million Americans watch birds around their homes (U.S. Department of the Interior 2009), 200,000 could be viewed as a drop in the bucket. Additionally, projects such as Celebrate Urban Birds aim for larger audiences of people who may not even consider themselves bird-watchers. Given limitations in staff resources needed to proliferate and manage publicity campaigns, viral online communications hold significant promise for increasing recruitment quickly and cheaply.

In 2008, someone unaffiliated with the Cornell Lab posted the availability of Celebrate Urban Birds kits on a website listing freebies. Within days, the project was flooded with requests for thousands of kits. Although staff were delighted with the sudden increase in requests, they were deeply concerned about their ability to meet demand: it costs about $1.75 per kit to cover instructions, data sheet, stickers, poster, and sunflower seeds for planting, plus postage and handling. It was also unclear whether people receiving the kits would use the freebies as intended. Future marketing efforts could test whether offering a more cost-effective freebie would result in a similar volume of requests and lead to active participation in the project.

Another viral communications strategy involves asking participants to use the same kinds of online tools that many fund-raising organizations use so people can ask their friends to pledge money for walkathons and other events. This method provides a vehicle for participants to involve people who are closest to them and gain recognition for the new participants they recruit to the cause.

Social networking also holds promise for building online citizen science communities as well as spreading news virally. The Cornell Lab of Ornithology has a Facebook page with more than 50,000 members that has grown organically with minimal staff effort devoted to it. Successful social networking sites require someone to tend and nurture them actively, and so far, their potential is largely unrealized for citizen science promotion.

On-the-ground community involvement also holds tremendous promise for the future growth of citizen science projects. Celebrate Urban Birds and the Great Backyard Bird Count owe their success to volunteer ambassadors and organizations that make the project their own through local events and activities. This model could also work well for projects such as eBird, which require a greater comfort with Web interfaces and bird-identification skills. These barriers can be brought down more easily in person than online. The Cornell Lab is investigating the possibility of connecting local eBirders with National Wildlife Refuges or nature centers. These volunteer ambassadors could lead bird walks and show people how to record and enter their data for the project. Although this effort would require additional investment to coordinate, the Celebrate Urban Birds model shows how community involvement creates a synergistic win-win for scientists and participants.

With effective promotion and community building, citizen science projects take on a life of their own, their momentum carried not only by promotional campaigns delivered by the institution but by the participants themselves. In 2010, the Cornell Lab of Ornithology and Natural Resources Defense Council launched an online community for conservation-minded bird enthusiasts at www.welovebirds.org. Community members who have no professional affiliation with the Cornell Lab often post information about eBird, Project FeederWatch, the Great Backyard Bird Count, and other citizen science projects. This microcosm within the birding community reflects a larger trend across the Internet whereby citizen science participants are spreading the word not only to their neighbors but to people across the country on blogs, social networking sites, and Listservs—helping nonprofits with limited marketing budgets to reach audiences on a larger and more frequent scale than ever before.

Acknowledgments
We thank our many colleagues who provided information and insights for this chapter, including Steve Kelling, Chris Wood, Brian Sullivan, Marshall Iliff, Karen Purcell, David Bonter, Tina Phillips, Janis Dickinson, and Rick Bonney. We also thank our colleagues who helped us try, learn from, and implement new approaches, especially Mary Guthrie, Jennifer Smith, Rob Fergus, Paul Green, Greg Delisle, Anne Marie Johnson, Genna Knight, Christianne White, Kathleen Gifford, and Kenyon Stratton. Finally, we give our heartfelt thanks to our partners, sponsors, volunteers, ambassadors, and participants who have given the citizen science projects a life of their own.

5

What Is Our Impact?

Toward a Unified Framework for Evaluating
Outcomes of Citizen Science Participation

TINA PHILLIPS, RICK BONNEY, AND JENNIFER L. SHIRK

Citizen science is a methodology that engages the public in large-scale scientific research while also attempting to achieve social and educational objectives. Despite the success of citizen science at answering biological questions at unprecedented scales (Bhattacharjee 2005; Cooper et al. 2007; Greenwood 2007b), there is growing recognition that the power of citizen science to promote science learning in informal environments has not effectively been measured or demonstrated (Bonney, Ballard, et al. 2009; Bonney, Cooper, et al. 2009; National Research Council 2009). Documenting educational and social impacts of citizen science is challenging because it is a highly interdisciplinary and relatively new field of study that has lacked a cohesive effort to develop criteria for defining and measuring success across projects. As a result, evaluations have tended to focus on isolated studies examining discrete educational outcomes of individual projects (Bonney 2004; Brossard et al. 2005; Evans et al. 2005; Krasny and Bonney 2005; Phillips et al. 2006; Trumbull et al. 2000).

Understanding the full breadth of outcomes resulting from project participation (e.g., gains in science process and content knowledge, increases in engagement and interest, changes in attitude and behavior) is important for several reasons. First, given the millions of dollars currently spent each year to develop and improve science education programs, funders typically require that plans for documenting broad impacts be included in project development. Second, understanding outcomes is necessary to develop a theoretical framework for guiding future research on the impact of citizen science on participants and society and improving methodologies for systematically studying this growing field. Third, information on outcomes is extremely valuable to professionals involved with educational aspects of citizen science who are looking to improve existing projects, develop new projects, and reach new audiences. Fourth, findings from outcome studies will improve our ability to set reasonable goals and objectives and suggest more evidence-based strategies for project development. Finally, compiling evidence concerning which approaches achieve maximum impact can increase our ability to replicate efforts across broad segments of the population.

We begin this chapter by discussing the historic rationale for evidence-based educational research. We next present a working matrix, modified from an evaluation framework developed under the auspices of the National Science Foundation (NSF) (Friedman 2008) for assessing educational impacts of citizen science and other public participation in scientific research projects (Bonney, Ballard, et al. 2009). We provide theoretical background on the elements of the matrix and demonstrate its use with an example from the Cornell Lab of Ornithology's NestWatch program. We conclude with recommendations for developing systematic evaluations of citizen science and other forms of public participation in scientific research.

Although citizen science can take place in classrooms (see Chapter 12), here we focus our attention on evaluation of citizen science in informal settings.

Background: Setting the Stage for Evaluation in Citizen Science

The modern era of U.S. science education began with the Soviet launch of Sputnik in 1957. This event inspired a new national agenda aimed at improving science literacy for the purpose of increasing global scientific competitiveness, advancing economic well-being, improving national security, and generally improving the public's ability to make informed science and policy decisions (Gregory and Miller 1998; Laugksch 2000; Rutherford and Ahlgren 1989). Increased federal support of the National Science Foundation led to school-based reform efforts such as the National Science Education Standards (National Research Council 1996) and Benchmarks for Science Literacy (American Association for the Advancement of Science 1993). Recognizing that Americans spend only a small percentage of their time in school, however, a national movement to foster lifelong learning of science outside of schools also emerged (National Research Council 2009), resulting in the 1984 establishment of the Informal Science Education (ISE) directorate within the National Science Foundation (National Science Foundation 2009). ISE is a broad field that aims to provide public access to and engagement with science via museums, science centers, television, radio, books, magazines, after-school programs, community-based services, and more recently, the Internet (National Research Council 2009). Sometimes referred to as "free-choice learning," ISE is voluntary, open-ended, and often unstructured learning that happens outside of school (Falk and Dierking 2002).

Many citizen science projects are developed within the context of ISE programming. For example, citizen science projects at the Cornell Lab of Ornithology are not only intended to gather large quantities of data, they also are designed to increase participants' knowledge of science content, to engage participants in the scientific process, to encourage positive attitudes toward the natural world, and to promote environmental stewardship practices. Of the 117 projects that have registered at the Citizen Science Toolkit website (www.citizenscience.org) since 2007, project managers report a three-way emphasis on educational, stewardship, and research priorities.

At the same time, the proliferation of ISE initiatives—including citizen science—has not kept pace with development of reliable methodologies to evaluate their educational effectiveness. As a result, the NSF has renewed its emphasis on increasing and improving program evaluation, particularly summative evaluation designed to assess a program's overall effectiveness. For the purpose of this chapter, we adopt the definition of program evaluation supplied by Trochim (2006): "*Evaluation is the systematic acquisition and assessment of information to provide useful feedback about some object.*"

But how exactly should projects be evaluated? What indicators of learning should be examined, and what techniques should be used? In an effort to address these questions, the National Research Council (2002) put forth a set of principles for accumulating knowledge and improving the quality of formal and informal educational research and evaluation through the use of evidence-based research. These principles rely on the use of rigorous, systematic, and objective procedures that can be investigated empirically, take into account alternative hypotheses, are capable of replication and generalization across studies, link research to theory, use valid and reliable measurements, and use experimental or quasi-experimental designs.

Addressing the ability of the United States to effectively compete in STEM fields (science, technology, engineering, and mathematics), a report from the Academic Competitiveness Council (U.S. Department of Education 2007) goes further to tackle the issue of scientific rigor in educational research by proposing a hierarchy of study designs for evaluating the effectiveness of STEM interventions. According to the ACC, evaluations should, whenever possible, take the form of experimental randomized control trials. When such rigorous experiments are not feasible, quasi-experimental studies utilizing well-matched comparison groups are considered the next-best option.

In reality, randomized control trials are seldom, if ever, implemented in informal settings (Brody et al. 2008; National Research Council 2009). This is mostly because designing such studies for citizen science and other free-choice learning interventions is fraught with challenge. The greatest difficulty is the fact that the majority of people who participate in citizen science in informal settings are self-selected, thereby eliminating the potential for random assignment to treatment groups. Similarly, obtaining unbiased control groups can be difficult.

Other challenges to designing evaluations that meet the ACC definition of rigor are ubiquitous throughout the ISE field. For example, audience diversity, while a programmatic asset, complicates program evaluation because of differences in the cultural, economic, social, ethnic, and geographical makeup of participants (National Research Council 2009; Rowe and Frewer 2005). Another issue is the problem of multiple influences on learning, that is, the everyday facets of life such as television, the Internet, and social interactions that influence individual learning and experiences. In other words, people tend to learn more over time no matter what educational programs they encounter; thus pre- and post-participation comparisons may produce evidence of learning that has nothing to do with actual engagement in the project. Because multiple influences are difficult to detect, they decrease our ability to isolate cause-and-effect relationships in noncontrolled environments.

The multidisciplinary nature of citizen science also presents unique evaluation challenges. Because projects may assemble experts from assorted, sometimes disconnected, fields (e.g., ecology, education, biology, communications, sociology, psychology), evaluations often have little or no coherent theoretical framework (Bonney, Ballard, et al. 2009; National Research Council 2009). Most of the theoretical constructs that are employed to evaluate public participation in scientific research tend to focus on policy-driven projects that utilize theories of participation, governance, and risk perception (Backstrand 2003; Irwin 1995; Lawrence 2006). The utility of these theories, however, is limited largely to the subset of participatory projects focused on environmental health or policy or both. Some scholars have explored the relevance of social learning (Fernandez-Gimenez et al. 2008) and social capital, i.e., the idea that social networks have value to groups and individuals (Overdevest et al. 2004) in participatory projects. Each of these, however, seems inadequate for describing either the breadth of project types or the complex interdisciplinary interests that drive the design of these projects and dictate the desired outcomes that evaluation must assess.

Yet another obstacle is the potentially high cost of comprehensive evaluations, even for small-scale projects. Furthermore, the ISE field lacks validated standardized assessments such as those that have been in place for decades within the formal educational system, making large-scale group comparisons difficult to achieve (National Research Council 2009).

Despite these challenges, citizen science practitioners should not avoid pursuing well-designed studies to measure intended (and unintended) learning outcomes aligned to programmatic goals and objectives. Evaluations should use the most rigorous study designs possible based on project needs, resources, and intended outcomes (Friedman 2008). Because most citizen science projects operate within a similar structure, it is feasible to design a common system of evaluating outcomes. In the following section we describe a framework developed in 2008 that can be used by citizen science projects to assess common categories of informal science education outcomes.

Measuring Educational Outcomes of Citizen Science

It is beyond the scope of this chapter to provide substantive instruction on how to design and conduct a summative evaluation. For this, we refer readers to many excellent resources including the *User-Friendly Handbook for Mixed Method Evaluations* (Frechtling and Sharp 1997); the *User-Friendly Handbook for Project Evaluation* (Westat 2002); *Evaluation: A Systematic Approach* (Rossi et al. 2004); and *The Practical Evaluation Guide* (Diamond 1999). However, in the next few pages we do present ideas to help guide overall thinking about evaluation design and instruments.

To maximize evaluation rigor, we suggest the use of a backward research design to first determine the desired outcomes for a particular audience and then to determine the best type of project to achieve those outcomes (Friedman 2008). Once the intended outcomes, stakeholders, audiences, and broad impacts are spelled out, project designers can work in concert with project evaluators to develop a logic model. This model helps project designers graphically articulate programmatic objectives and strategies and allows evaluators to focus their attention on key interventions and intended outcomes (see W.K. Kellogg Foundation 2004 for more information on developing a logic model).

To demonstrate the logic model approach, a generic citizen science logic model, sometimes referred to as a program theory, is shown in Figure 5.1. The logic model is based on the Outcome Approach Model supplied by the W.K. Kellogg Foundation (2004) and illustrates generalized volunteer *activities* and *outcomes*, which can then be adapted to a specific project's contexts and needs. The *inputs*, or resources brought to a project, include the elements necessary for volunteer participation, such as personnel, funding, and the infrastructure for recruitment and training. In most large-scale citizen science projects, *activities* focus on the collection and reporting of data, but some projects provide opportunities for participants to engage in other aspects of the research process. The *outputs* typically demonstrate short-term results of activities, for example, an amount of data collected by

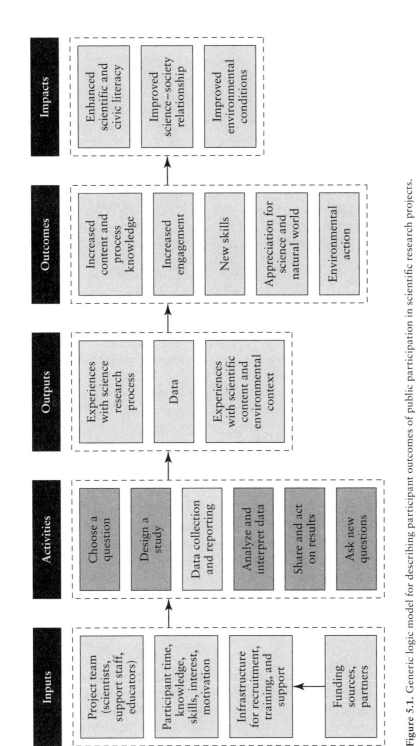

Figure 5.1. Generic logic model for describing participant outcomes of public participation in scientific research projects.

participants or the lived experience of collecting the data. The *outcomes* are the short- and long-term objectives that the program hopes to achieve and that the evaluation usually attempts to measure. *Impacts* of large-scale citizen science projects are generally long-term, broad in scope, aimed at expanding knowledge and capacity for a particular field of study, and meant to provide benefits to society. Although they are more difficult to measure, impacts are important to describe and are of particular interest to funding agencies.

An important resource for developing evaluation plans is the *Framework for Evaluating Impacts of Informal Science Education Projects* (Friedman 2008). Written partly as a response to the ACC report of 2007, the *Framework* provides suggested evaluation designs and a standard set of categories that allow ISE programs to collect project-level outcomes in a systematic way and make cross-project and cross-technique comparisons of outcomes. A major contribution of the framework is the formulation and description of agreed-on outcomes (referred to as *impact categories*) common among ISE programs. The six major categories include knowledge, awareness, and understanding; engagement and interest; attitudes; skills; behavior; and "other." Given the knowledge gaps previously mentioned, the "other" category allows for the inclusion of emergent outcomes in unanticipated aspects of educational impact.

Although not specifically designed for evaluation of citizen science projects, the framework has been used by a team of educational researchers to develop a more refined tool for assessing a series of "public participation in scientific research" (PPSR) projects (Bonney, Ballard, et al. 2009). The researchers developed subcategories for each of the framework's six impact categories to capture finer-level outcomes. The result was a matrix that could be used to document goals, outcomes, and potential indicators for each impact subcategory (Table 5.1). The matrix is available online at www.citizenscience.org with sample goals and indicators included for each subcategory. Development of the matrix is ongoing as new subcategories are created, instruments assessed, and new knowledge generated. It is important to note that no single project can influence or evaluate all or even most of the stated goals within the matrix.

Application of the Matrix

The Program Development and Evaluation group at the Cornell Lab of Ornithology has begun using this matrix to provide guidance in project development and assessment for Cornell Lab of Ornithology projects as well as those based at other institutions. The process starts in the project development phase as evaluators and program staff refine vague goals such as "to increase science literacy" by determining a set of desired outcomes

TABLE 5.1.
Application of the public participation in scientific research matrix to NestWatch

	Sample goals	Potential outcomes	Sample indicators
Framework impact category: awareness, knowledge, understanding[a]			
Matrix subcategory *Science content, concepts* Short- or long-term understanding of a STEM-related topic, principle, phenomenon, or theory	Increase public understanding of nesting biology	a. Participants understand that bird mating systems are highly variable b. Participants understand the avian nesting cycle, from nest building to fledging c. Participants understand how the nesting cycle differs between altricial and precocial birds d. Participants understand that natural variability of breeding parameters occurs, especially across species, time, and space	Increase in correct responses from pre- to post-project knowledge survey Analysis of written comments on data-entry forms, open-ended survey questions, e-mail discussions
Framework impact category: engagement or interest[b]			
Matrix subcategory *Project/activity* Interest in and pursuit of, further activities related to the project	Increased quality and quantity of participation with NestWatch	a. Increased number of new participants and increased time spent engaging with the project b. Increased dialogue between participants and scientists	Retention and recruitment rates, numbers of nest records submitted annually, frequency of data submission

Notes: NestWatch is a citizen science project of the Cornell Lab of Ornithology. The full matrix is available online at www.citizenscience.org. Definitions for impact categories were adapted from Friedman 2008.

[a] Measurable demonstration of, assessment of, change in, or exercise of awareness, knowledge, and understanding of a particular scientific topic, concept, phenomenon, theory, or career central to the project

[b] Measurable demonstration of, assessment of, change in, or exercise of engagement or interest in a particular scientific topic, concept, phenomenon, theory, or career central to the project

whose achievement can be observed given the time and resources that the project has or will have available. Measurable outcomes refer to the specific, measurable, attainable, realistic, and timely objectives that the project hopes to achieve and that can be empirically recorded (Doran et al. 1981). Developing indicators to assess how well a project achieves its objectives is useful for establishing how success will be defined.

To illustrate how to use the matrix to measure selected outcomes, we provide examples from an evaluation of NestWatch (www.nestwatch.org), an NSF-funded citizen science project developed by the Cornell Lab of Ornithology. Results from this evaluation are currently in preparation. We also provide some theoretical groundwork behind selected matrix subcategories.

NestWatch is designed to engage volunteers in monitoring bird nests, collecting data about eggs, nests, and nestlings, and submitting their observations to an online database. Scientists use the data to study the relationship between environmental change and the nesting success of birds breeding across North America. The educational goals of NestWatch include (1) increase public understanding of nesting biology, (2) increase public engagement with the scientific community, (3) improve participant nest-monitoring skills, (4) increase appreciation for the natural world, and (5) increase environmental stewardship practices of participants that benefit breeding birds. Here, we describe each of these goals within the context of the framework and matrix.

Awareness, Knowledge, and Understanding

Included within the framework impact category "awareness, knowledge, and understanding" are several matrix subcategories such as knowledge and understanding of science content, science process, the nature of science, and careers in science. Although it is tempting to lump all of these into one general category of science literacy, there is little agreement on a concise definition (Bybee 1995; Hurd 1998; Lederman 1998; Miller 1983; National Research Council 1996, 2009), and some even question its practical attainment by society (Bauer 1992; Shamos 1995). Many studies have been conducted that attempt to measure scientific literacy based on knowledge about specific content, typically of the sort found in textbooks, with disappointing results (Durant et al. 1989; Evans and Durant 1995; Miller 1983, 1998). According to DeBoer (2000), rather than trying to develop a standard way of measuring the many facets of science literacy, a better measure is a public that is interested and engaged with science issues and able to apply science to individual circumstances and social interactions. In this vein, we argue that projects should attempt to measure only those aspects of science literacy that they can reasonably influence and that are relevant to project participation.

The first subcategory, increasing science content knowledge, is one that most citizen science projects can reasonably achieve, especially if goals are relevant to specific content such as plant phenology, breeding biology, or water quality. Provided that project activities are truly aligned with program content, project developers and evaluators should work in tandem to create project-specific items that can be used to assess content-specific learning outcomes.

In the case of NestWatch, project staff articulated four potential indicators for the first content knowledge goal, "Increase public understanding of nesting biology." Because we were interested in measuring demonstrated change in knowledge rather than cause and effect, we developed a matched pre- and post-project survey that asked for information that would demonstrate knowledge gains specific to breeding biology.

Engagement

By their very nature, citizen science projects increase public engagement with science, although the level of engagement can vary among projects. Matrix subcategories include engagement with science content, process, project, and careers. Measuring outcomes related to engagement can be fairly straightforward, particularly if evaluators use quantitative methods such as the number of annual participants, retention/recruitment rates, amount of data collected, number of people attending science conferences, and number of people showing heightened interest in science careers. Frequently used metrics to assess science interest and engagement also can be adapted from the National Science Board Indicators (National Science Board 2002).

To evaluate engagement with the NestWatch project we measured the number of annual participants, retention and recruitment rates, numbers of nest records submitted annually, and the frequency with which participants submitted data. We also examined the frequency of online exchanges between participants and scientists through analysis of e-mails, discussion forums, Listservs, and social networking sites.

Skills

Citizen science projects can significantly influence acquisition of science-related skills. Most of the matrix subcategories focused on skills are similar to those involved in using the "scientific method" (i.e., asking questions; designing studies; collecting, analyzing, and interpreting data; discussing and disseminating results). The matrix also includes skills related to the use of technology and communication. The skills that citizen science projects can most effectively influence are related to data collection and include the ability to collect specimens or samples, identify organisms, use measurement instruments, follow protocols, and use proper sampling techniques consistently over time. Project developers and evaluators often rely on quality control and data assurance measures such as adherence to protocol or demonstrated mastery of a technique to determine skill acquisition. In small-scale projects, evaluating participants' abilities to conduct these skills during workshops has proven valuable (Wilderman et al. 2005). Direct observation of skills in large-scale projects can also be implemented at least within a sample of the population, but logistical issues regarding time and resources must be taken into account.

In the NestWatch example, the goal "Improve participant nest monitoring skills" was examined in two ways. First, all NestWatch participants were encouraged to take an online nest monitor certification quiz that assessed their understanding of proper nest-monitoring techniques. We were able to track their score each time they took the quiz, thereby documenting changes in their understanding of proper techniques. Second, by looking at the data submitted by individual volunteers, we determined the proportion of volunteers who changed the way in which they monitored nests (i.e., the frequency of nest checks) and how they reported their observations (i.e., as a simple nest summary or as a more informative chronological account of each nest visit).

Attitudes

Attitudes are often described as a three-pronged entity made up of cognitive (thoughts and ideas), affective (feelings and emotions), and behavioral (actions and intentions) components (Aiken 2002). The literature base on attitudes toward science is vast (Bauer et al. 2000; Durant et al. 1989; Evans and Durant 1995; Miller 2004; National Science Board 2002; Sturgis and Allum 2004). Yet despite the academic focus, research on attitudes toward science is still muddled, in part because of a lack of a clear definition (Aiken and Aiken 1969) and a lack of valid and reliable instruments (Haladyna and Shaughnessy 1982; Pearl 1974).

The complexity of measuring attitude structures is exemplified by the shortage of studies that document attitude change as a result of citizen science participation. Given these complexities, we suggest using one of the numerous existing instruments such as the Attitude toward Organized Science Scale (National Science Board 1996) or questions from the Science and Engineering Indicators (National Science Board 2002) for comparisons to a larger population. Also, because many citizen science projects have an environmental focus, instruments that measure attitudes toward nature and the environment also can be used, including, for example, the New Environmental Paradigm (Dunlap and Van Liere 1978); the New Ecological Paradigm (Dunlap et al. 2000); the Environmental Identity Scale (Clayton 1993); and the Nature Relatedness Scale (Nisbet et al. 2009).

The NestWatch goal to "increase appreciation for the natural world" aligned well with the attitudes-toward-environment subcategory and was assessed using the preexisting Environmental Identity Scale (Clayton 1993), which has been widely used to measure self-identification with nature.

Behaviors

Like the problem of measuring attitudes, measuring behaviors is a complex endeavor because of the interactions among behaviors and knowledge, attitude, and intention. By all accounts, however, documenting behavioral change is considered the "holy grail" of impact studies. Much of what we understand about human behavior is derived from the psychological literature (see, for example, Ajzen 1985; Fishbein and Ajzen 1975; Schwartz 1977). More recent theories from environmental and conservation psychology fields are emerging that deal specifically with environmentally responsible behavior (Geller 1995; Kollmuss and Agyeman 2002; Stern 2000). Despite numerous theories, a universal approach to evaluating behavioral outcomes remains nonexistent, mostly because of the need for measuring specific behaviors. Consequently, project developers often assume that specific participation activities lead to more universal environmental behavior such as reducing carbon emissions, recycling, and conserving energy. These indirect cause-and-effect relationships are difficult to discern, and the tendency to lump similar behaviors "may be rooted in the notion that a generalized attitude toward the environment underlies all types of pro-environmental behavior" (Vining and Ebreo 2002:27). As a general rule, only those behaviors that are relevant to project activities should be measured. These include, for example, community involvement, citizen action, environmental stewardship, and new participation. Also, to increase accuracy of self-reported behaviors, Vining and Ebreo (2002) suggest obtaining quantifiable frequencies rather than simply asking for presence or absence of behaviors.

The NestWatch goal "Promote environmental practices that benefit breeding birds" falls into the "environmentally responsible behavior" subcategory and was measured using a pre- and post-participation questionnaire that quantified frequencies for behaviors such as removing invasive plants, eliminating pesticide use, providing nesting structures, and buying shade-grown coffee. It is important that each of these behaviors is closely linked to information provided by the project, making the connection between knowledge and behavior change realistic and feasible. In addition to the above-mentioned data sources, qualitative interviews and analyses of open-ended questions were conducted on a subsample of NestWatch participants to obtain richer explanations of measured impacts and to explore unintended outcomes. When feasible, study designs should incorporate a mixed-methods approach to achieve complementarity among approaches, obtain triangulation of findings from several different methodological sources, and increase the richness of the study (Greene et al. 1989; National Research Council 2002).

The NestWatch examples that we have provided illustrate how one project was able to use the matrix to guide summative evaluation. But note that although this ambitious project tried to measure achievement of outcomes within at least one subcategory of each major category, this is not necessary or even advisable for all citizen science projects. For example, a project may focus on skill acquisition and strive to achieve outcomes in several or all of the skill subcategories but none of the other major categories. The matrix was developed to be flexible and to provide a framework suitable for most projects irrespective of resources. In the end, practitioners should focus on specific outcomes that can be achieved through project participation rather than trying to check off all potential learning outcomes outlined by the framework and matrix.

※

Currently our understanding of the outcomes of engagement in citizen science and other public research initiatives is incomplete but advancing as relevant theories are uncovered and generated. To advance learning about impacts in a systematic way, we suggest using the matrix to guide outcome evaluations. We emphasize that although formal standardized instruments are available and can be used when appropriate, it is equally important to assess informal learning in a manner aligned with the nature of the project and to allow for evaluation outcomes that will break new ground. Whatever methodological techniques or instruments are used, evaluations should employ the most rigorous designs possible, present unbiased accurate results, and contribute to a knowledge base that informs this growing field.

To fully understand the practice of citizen science, however, future research must also focus on understanding the multiple dimensions of participation. Qualitative studies examining motivations for participation, informed by theories such as those offered by the volunteer literature (Clary and Snyder 1999; Clary et al. 1998; Lawrence 2006; Nerbonne and Nelson 2008), offer compelling points of departure to investigate motivations born from altruism, desire for continued learning, self-enhancement/satisfaction, career goals, empowerment, social involvement, and connection with nature. Additionally, new research suggests that individuals may be motivated by self-identification with science and scientists—a phenomenon that has yet to be fully described and understood (Falk and Storksdieck 2009; Fraser et al. 2009; National Research Council 2009).

Instrument development suitable for capture of outcomes related to participation in scientific research is another area of study that should be critically examined. Of particular interest are instruments that can be customized for specific topics or projects so that staff need not begin each evaluation from scratch. Commonly used instruments include the Attitude toward Organized Science Scale (ATOSS) used in National Science Board surveys (National Science Board 2002); Views on Science-Technology Society (Aikenhead and Ryan 1992); the Views of Nature of Science Questionnaire (Lederman et al. 2001); and the New Environmental Paradigm scale (Dunlap et al. 2000; Dunlap and Van Liere 1978). Some of these have been used in previous Cornell Lab of Ornithology project evaluations with disappointing results. Generally they are not fine grained enough to measure changes in individuals who are already literate in a given area. In addition, the scoring rubrics for some generalized instruments have been criticized for not recognizing the multiple meanings that science may have for audiences with different cultural backgrounds or levels of experience with scientific research (Bauer and Schoon 1993; Bauer et al. 2000; Collins and Pinch 1993; Godin and Gingras 2000; Kallerud and Ramberg 2002). Nevertheless, we believe that generalized instruments can be developed for citizen science and used by multiple projects to allow cross-project and cross-technique outcome comparisons.

Finally, the lack of a theoretical framework to guide future research on citizen science must be addressed. The logic model presented here is a first step toward developing theory. To facilitate continued development and codification of citizen science, however, there is need for further exploration of existing ideas such as activity theory (Roth and Lee 2002), communities of practice (Lave and Wenger 1991), extended peer communities coming out of discussions of post-normal science (Ravetz 1999), the coproduction of knowledge (Corburn 2003), and the recently proposed ecological framework (National Research Council 2009).

Acknowledgments

We thank the members of our CAISE Inquiry Group on documenting the impacts of Public Participation in Scientific Research Projects for their input into the matrix presented here: Heidi Ballard, Rebecca Jordan, Ellen McCallie, and Candie Wilderman. This work was supported by the National Science Foundation grant ESI-0610363. All opinions are those of the authors and not necessarily of the NSF.

PART II

IMPACTS OF CITIZEN SCIENCE ON CONSERVATION RESEARCH

This section of the book illustrates the questions, methods, and potential research impacts of citizen science with detailed discussion of a few projects. We focus on birds because large-scale projects with other taxa are too new to have accumulated a body of research sufficient to illustrate the full range of research possibilities. The major ecological questions addressed here, however, as well as discussions of sampling design, analysis methods, and tools, are applicable across a broad array of taxa.

Today, the Breeding Bird Survey weighs in with more than 500 publications and the Christmas Bird Count with more than 300. Although the broader community of researchers was initially skeptical of citizen science data, careful analysis rapidly increased appreciation of the scientific value of the data, so that citizen science results were beginning to appear in top scientific journals by the 1990s (Bhattacharjee 2005) and data analysis methods, rather than data quality, became the primary focus of critical reviews. The value of large, longitudinal, spatially expansive data sets is now widely accepted, and we are seeing rapidly growing interest in tackling spatial ecology and conservation questions at scales that encompass the entire ranges of plant and animal species. Such scales are necessary for understanding the impacts of widespread ecosystem perturbations including global climate change, the spread of new infectious diseases, ecological invasions, and changes in the quality or distribution of habitat across landscapes.

While the primary value to ecologists of large-scale citizen science data is the power of long time series and large spatial coverage to illuminate range-wide patterns, hypothesis testing is also possible by treating patterns revealed by the data as "initial observations" analogous to preliminary field observations in on-the-ground ecological studies. Where observations generate new hypotheses and predictions, these can be tested through additional data analysis or with experiments (Dickinson et al. 2010). Perhaps the best research model for citizen science is one in which citizen science monitoring is combined with localized experimental research, as this approach allows simultaneous understanding of large-scale ecological patterns and the mechanisms that underlie them.

The chapters in Part 2 highlight a mix of approaches to working with large-scale citizen science data to address scientific questions. These approaches include (1) using citizen science data and conventional statistical models to generate patterns as initial observations that then lead to more intensive, hypothesis-driven studies (Chapters 6, 7, 9, and 11); (2) using data mining as a novel means of discerning patterns and generating new hypotheses (Chapter 8); and (3) combining a priori experimental or observational studies with citizen science, either as side projects or by embedding them within citizen science projects to answer questions at a variety of scales (Chapters 6 and 9).

The chapters in this section can also be dissected, albeit imperfectly, by emphasis. In Chapter 6, Cooper et al. focus on the types of ecological questions that are best suited to citizen science; in Chapter 7, Zuckerberg and McGarigal explore the bridge between citizen science and landscape ecology; and in Chapter 8, Fink and Hochachka focus on data mining and exploratory analysis.

Part 2 ends with two chapters that use citizen science to conduct conservation research with explicit policy implications. In North America, citizen science data have been used to create land managers' guides, set conservation priorities through Partners in Flight or annual State of the Birds Reports, and, in the case of the Breeding Bird Survey, identify species at risk. In Chapter 9, Hames et al. (2002a) describe a specific case in which "super citizen scientists" used manipulative sampling to gather data implicating acid rain and mercury in forest bird declines, highlighting advantages of partnerships with governmental and nongovernmental organizations. In Chapter 10, Greenwood describes in detail the British Trust for Ornithology's (BTO) highly successful model for engaging volunteers in scientific research. BTO is publicly funded, which positions it to have impact, is driven by the participants, and is protected against the infiltration of nonscientific endeavors, such as advocacy. While it is not yet clear whether such a model could work as well in a less nature-focused culture or on a larger "island," we would be remiss not to provide a thorough description of the British citizen science perspective within this book.

6

The Opportunities and Challenges of Citizen Science as a Tool for Ecological Research

CAREN B. COOPER, WESLEY M. HOCHACHKA,
AND ANDRÉ A. DHONDT

In this chapter we explore the potential for large-scale citizen science to advance our understanding of ecological systems. We will (1) note the types of ecological research questions for which the scale (extent and resolution) of data from citizen science is particularly suitable, and (2) highlight research considerations that need to be taken into account when designing (or continuing) a citizen science project. We hope this chapter will encourage ecologists to see novel ways to make use of citizen science to examine large-scale patterns, effects of spatial variation on the processes they study, and higher levels of biological organization.

Within the term *citizen science*, *science* refers to the process of investigation rather than to the other meaning of *science* as a body of accumulated knowledge. The scientific process is often idealized as being fairly linear: researchers define a question based on prior knowledge or conjecture, make observations, form testable hypotheses, and conduct studies designed to potentially falsify these hypotheses (Popper 1959), leading to strong inference and publication of the results. The only criterion for scientific (as opposed to nonscientific) investigation is that the hypotheses have the potential to be falsifiable. Investigation is thus informed by a combination of theoretical models and carried out with observational tests of predictions, or, when possible, experimentation.

The collection and publication of natural history observations provides a base of knowledge from which initial questions and testable hypotheses can be formed. In many fields of science, avenues of research have progressed to the point where investigations based on collection of nonexperimental observations, such as natural history observations, are perceived as not rigorous enough to push frontiers. Our belief is that this is not true for many ecological questions, specifically because patterns and processes can vary among locations or through time: intensive local studies are not sufficient to understand all ecological processes. One unique aspect of citizen science is that it facilitates the gathering of the natural history observations at enormous spatial and temporal scales. Thus, we argue that if used properly, large-scale citizen science data can open new avenues in ecological research. Indeed, even deciphering patterns from basic natural history data collected at large spatial and temporal scales often requires novel and complex analytical techniques (see Chapter 8).

Large-scale patterns built from citizen science methods can be employed strategically to advance ecological research in two ways: (1) by providing aggregations of observations that reveal patterns at larger scales and/or higher levels of organization, thereby generating new lines of inquiry into the processes that might account for newly discerned patterns; and (2) by allowing researchers to test mechanistic hypotheses that predict large-scale, or higher-level, patterns. These two options mirror the dual ways that citizen science schemes are formed, either with protocols designed to collect data to address a specific research question or, more commonly, with protocols that are designed robustly within the broader goal of monitoring without specific a priori hypotheses.

Higher-Level Phenomena and the Utility of Citizen Science

Often the creation of a new tool is responsible for opening up a new field of investigation. For example, tools to color band birds expanded the field of behavioral ecology and molecular tools expanded the field of evolutionary ecology. Many of these new tools allow scientists to examine mechanisms at lower levels of biological organization. As explained above, citizen science creates new research opportunities by increasing the geographic scale over which observations are gathered to gain insights into processes and mechanisms at work at higher levels of biological organization. Citizen science is especially useful when it not only provides data collected over large spatial or temporal extents but also provides data collected at fine resolution to address underlying processes. Logistically, citizen science works best when data can be accurately and cost-effectively obtained with direct public involvement in (at least) the data-collection process and with data validation tools in place (Chapter 3).

The endeavor of research via the citizen science "tool" can encompass the use of data that have already been collected by a project, use of an existing network of participants to collect new data, or development of an entirely new project. In the next section, we provide examples to illustrate the potential of citizen science for ecological research by describing a suite of questions that require observations across a large geographic scale or during extended periods.

Opportunities for Research

Processes Underlying Patterns of Distribution and Abundance

Ecology is the study of the distribution and abundance of organisms (Andrewartha and Birch 1954; Krebs 2009) and their interactions with the environment. Citizen science data have been used extensively to describe species' distributions (Gaston 2003). The processes underlying these patterns can also be elucidated using citizen science data, as the following examples illustrate.

Identifying Habitat Associations

As noted in Chapter 7, some forms of data that are gathered over large scales, such as remote-sensing data, can be easily collected without the need for citizen scientists. In most cases, however, data on the presence of species at a particular site can be collected only by human observers (Kelling et al. 2009). Therefore, a marriage between citizen science and remote-sensing data is often required to answer questions about the relationship between environmental variables and the distribution, abundance, phenology, or behavior of animals (or plants). For purposes of conservation and management, habitat associations can be determined with this combination of data (e.g., Brotons et al. 2004; Kéry et al. 2005; Chapter 9). Macroecology (Brown and Maurer 1989) is another facet of ecology where this marriage is fruitful.

Assessing the Impacts of Environmental Change

The accumulation of observations over long time scales has great utility for tracking changes in abundance, and this is best exemplified by the use of citizen science data from the North American Breeding Bird Survey, which has served as a key ingredient for conservation planning in North America, both for early identification of declining species and as the basis of detailed regional plans for conservation action (Rich et al. 2004). Citizen science data have also contributed significantly to our understanding of the effects of climate change on the ranges of species and their reproductive ecology. Thomas and Lennon (1999) used data from breeding bird atlas schemes in Great Britain to show that the breeding ranges of birds were expanding northward, and Hitch and Leberg (2007) found similar results in North America based on data from the Breeding Bird Survey. Citizen science data have shown similar northward shifts for other taxa, such as dragonflies in Britain (Hickling et al. 2005). It is important to note that none of the data used to provide these insights into the biological impacts of climate change came from schemes designed with protocols for that purpose. These examples illustrate that researchers can repurpose long-term data to address a myriad of questions, or as Dhondt (2007) put it, "They provided answers before the questions were asked."

Urban ecology is a specialized approach to understanding impacts of environmental change and how urban environmental practices might play a role in conservation. Although researchers may typically view remote areas as the most difficult to access for data collection, making observations around residences also can be logistically challenging. Engaging residents in data collection via citizen science allows easy sampling of private property, which usually takes up the majority of land in urban environments. For example, in Project Squirrel, participants count Fox and Gray Squirrels around their residences (van der Merwe et al. 2005). Based on Project Squirrel data in the Chicago area, Fox Squirrels were found to be more abundant than Gray Squirrels in areas with single-family homes and in association with elms and maples, as well as in areas with cats and dogs. In contrast, Gray Squirrels were associated with oaks and pines, multifamily homes, high-rises, and areas with fewer cats and dogs (van der Merwe et al. 2005). In Neighborhood Nestwatch, administered by the Smithsonian Migratory Bird Research Center, researchers coordinate their work around residences with participants at those residences, thus facilitating access and increasing observer effort by involving the people who are more frequently present at the sites, who then continue to make observations.

Dispersal, Migration, and Movement

Although migratory movements of animals are challenging to track, the multiple eyes and ears of citizen scientists can make tracking large-scale movements feasible. For example, the Vanessa Migration Project (www.public.iastate.edu/~mariposa/Vanessaproject.htm), organized in partnership with Iowa State University Geographic Information Systems Facility, Iowa Nature Mapping, and the Iowa Environmental Mesonet, tracks movement and outbreaks of butterflies. Bird migration is traceable through eBird and networks of migration monitoring stations such as the Canadian Migration Monitoring Network (www.bsc-eoc.org/national/cmmn.html) and the North American Hawk Migration Association (www.hmana.org). Any project that collects data on marked individuals or provides data on temporal changes in distribution and abundance within a year can provide potentially new information on animal movement.

Emerging Diseases

Disease-causing organisms are invading species that can have particularly important influences on ecosystems, but the emergence of new diseases is challenging to track. One of the best-studied emerging pathogens in wild animal populations is the bacterial pathogen *Mycoplasma gallisepticum,* which first appeared causing disease in House Finches in the mid-1990s in eastern North America. Dhondt et al. (2005) tracked the epidemic spread of mycoplasmal conjunctivitis in House Finches through a citizen science program, the House Finch Disease Survey (HFDS) administered by the Cornell Lab of Ornithology. Participants in the HFDS collected and reported data that were used to index disease prevalence (Altizer et al. 2004) and its impact on host abundance (Hochachka and Dhondt 2000). The HFDS was able to start within 9 months of the first reports of disease, because the majority of initial participants in the survey were recruited from the ongoing Project FeederWatch (Chapter 2). In this case, a citizen science project designed for monitoring winter birds served as the springboard for another project with much more specific goals, show-ing the potential ability of long-term data-collection programs to quickly gather supplemental information when unique circumstances arise. Simi-larly, existing monitoring projects were harnessed to obtain information critical to making predictions for how and where avian flu might reach the UK (see Chapter 10).

Geographic and Temporal Trends in Life History and Behavior

In addition to understanding processes underlying changes in distribution and abundance, citizen science can be used to better understand facets of life histories of organisms; for example, citizen science data can be used to examine differences in details such as demographics, traits, and phenology. Again, the strengths of using citizen science to investigate these facets of life histories is the ability to gather ecological data over a larger area than would otherwise be feasible in order to identify and understand variation in patterns across space or through time.

For example, advancements in laying dates varied across populations of tits (*Parus* sp.) across Europe (Visser 2003). At northern sites that have not experienced warming, and in habitats at southern sites where phenology of food sources had not changed, Great and Blue Tits did not show changes in laying dates. At intermediate latitudes, populations that historically lacked second clutches showed advances in laying dates, while populations with second clutches showed a reduction in the frequency of second clutches with a smaller advance in laying date (Visser 2003). Relatively few studies of this type exist, so much of our understanding of large-scale and long-term variation in life histories has come from citizen science data. Below are examples of past uses of citizen science data, or suggestions for fields in which the use of citizen science data could provide more detailed informa-tion on large-scale variation in life history traits.

Demography and Life History Traits

Further understanding of geographic variation in reproductive rates will almost certainly require the use of citizen science data. Initial work in this field (Lack 1947; Moreau 1944) described gradients in clutch sizes between tropical and temperate birds, across seasons, between islands and continents, across continents, and with changes in elevation. These early efforts to examine geographic patterns in life history traits were coarse, with a few exceptions (Johnston 1954), comparing tropical (equatorial) and temperate (higher latitude) birds (Lack 1947; Moreau 1944; Skutch 1949), sometimes with entire countries comprising a single data point (Lack 1943). Nevertheless, the results from such studies provided the foundation for development of life history theory in ornithology (Ricklefs 2000). Although the resolution of these large-scale studies has improved over time as more locations have been added (Young 1994), citizen science projects can still provide additional data to deepen our understanding of geographic variation in reproductive strategies.

Data from citizen science projects have already contributed to our knowledge in this realm. For example, Peakall (1970) presented an analysis, similar to that of Lack (1943; Figure 6.1), of data on clutch size of Eastern Bluebirds using over 8000 records from Cornell's first citizen science project, the Nest Record Card scheme. He was able to show, averaging clutch sizes within states, that Eastern Bluebird peak clutch size varied geographically (slightly higher in the center of the range at the height of the breeding season), but without an obvious latitudinal trend (Figure 6.2). The reason for this apparent lack of latitudinal trend was finally elucidated when Dhondt et al. (2002) examined the data further, and found that at southern latitudes, clutch size was low at the start of the season, which began earlier than at central and northern locales (Figure 6.3). In all areas, clutch size decreased from late April through July. These temporal trends masked a small but clear latitudinal increase in clutch size to the north.

Such complex variation in reproductive patterns over broad geographic areas likely can be detected only using citizen science data, and so far we have just scratched the surface in using such data to examine patterns such as seasonal trends across latitudes in clutch size (Cooper, Hochachka, Butcher, et al. 2005; Dhondt et al. 2002), reproductive effort (Cooper, Hochachka, and Dhondt 2005), and hatching failure (Cooper et al. 2006).

Seasonal timing of life events is important to measure as temporal shifts can provide sensitive indicators of response to environmental conditions. Citizen science data collection is well suited for collecting the information to study phenological patterns. Changes in phenology are predicted in a diversity of taxa as a response to climate change, and several have been examined using citizen science data. Data from butterfly monitor-

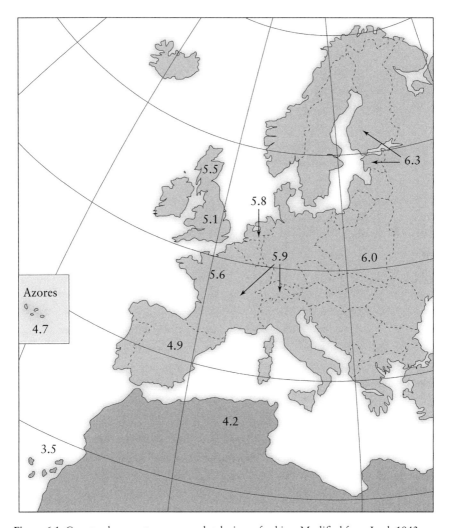

Figure 6.1. Country-by-country average clutch sizes of robins. Modified from Lack 1943.

ing schemes in Britain have shown a tendency toward earlier first appearances in the late twentieth century (Roy and Sparks 2000). Timing of birds' breeding seasons has also been found to change; for example, Dunn and Winkler (1999) used nest record card data to discover that Tree Swallows had advanced their laying dates by up to 9 days in response to changing climate. Also of concern are changes in plant phenology. Project BudBurst (Chapter 2) monitors geographical and temporal variation in plant phenology as gateway project to a more intensive phenological monitoring scheme launched by the National Phenology Network in 2010.

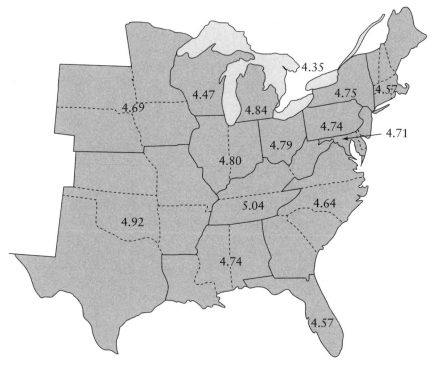

Figure 6.2. Eastern Bluebird, *Sialia sialis*, average clutch size at the peak of the breeding season for fifteen regions. Modified from Peakall 1970.

Geographic and Temporal Trends in Behavioral Traits

While citizen science projects that involve counts or recording presence/ absence may be easiest to implement because they tap into what many hobbyists already do, citizen scientists can also be asked to provide more detailed data using specific protocols for assessing behavior, such as the breeding evidence codes used in breeding bird atlases, or using specialized instruments to collect additional types of data that extend beyond simple observation. Both direct and indirect observation can be used to study behavior. At the Cornell Lab of Ornithology, participants in Project PigeonWatch follow a protocol and gather data on mate choice behavior in pigeons across the globe to examine the potential role of mate choice in maintaining variation in color morphs. We have also created projects in which citizen scientists observe and classify behaviors recorded on camera using the CamClickr project's drag-and-drop method of sorting still images by behavior or state (Voss and Cooper 2010).

We have successfully worked with project participants to record behaviors without observing the birds at all; participants placed temperature-recording data loggers into nests, where the temperature records were then used to infer incubation rhythms of the birds (Cooper and Mills 2005).

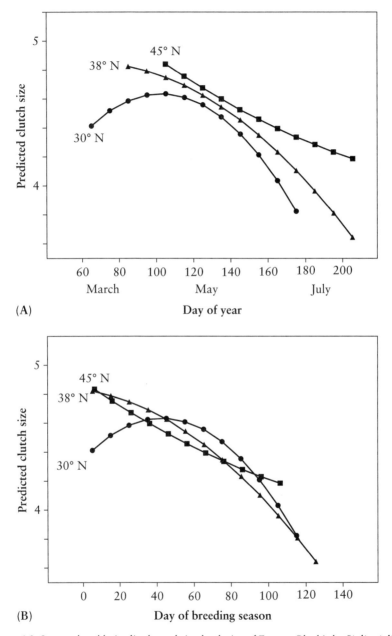

(A)

(B)

Figure 6.3. Seasonal and latitudinal trends in clutch size of Eastern Bluebirds, *Sialia sialis*, compared relative to (A) calendar date (day of year), and (B) the start of the breeding season. Modified from Dhondt et al. 2002.

Perhaps the most important point to draw from these examples is that creative use of technology can expand the range of types of studies that are amenable to a citizen science approach.

All of the examples presented above show that ecologists can study diverse ecological patterns and processes over large areas and long time periods. Citizen science data, however, clearly should be viewed as complementary to data from other sources if the goal is to build a full understanding of ecological processes (Dhondt et al. 2005). They can also be used at different stages in the progress of a research program; Chapter 8 discusses this in detail, talking about the use of citizen science data to generate hypotheses for subsequent testing in an early, exploratory phase of research (see also Dickinson et al. 2010).

Regardless of their specific uses, citizen science data need to come from studies designed to gather relevant data that are appropriate for rigorous analysis. Design constraints and challenges are common to any research protocol, but our experience has shown us that there are several challenges that are particular to, or strongly manifested in, citizen science projects.

Rare Observations and Occurrences

Rare events and events that are rarely detected can be very important but are difficult to study. The cumulative efforts of dispersed networks of citizen scientists can contribute an army's worth of person-hours of work, making the collection of information on rare events tractable. Examples of such information or events include predation events (witnessing predation), mortality events (discovering dead individuals), and sightings of rare or elusive species.

Observations and Experiments Involving Rare Events

A key example of the power of citizen science to study rare ecological phenomena is with invasive species or disappearing species. Rare species can be located by project participants, as has been seen with The Lost Ladybug Project (www.lostladybug.org), administered at Cornell University, in which digital images of various species of ladybugs are received from thousands of locations across the United States (Losey et al. 2007). Through this project, two children in Virginia reported a nine-spotted ladybug, which researchers had not documented in 14 years (Losey et al. 2007), and a 6-year-old in Oregon was able to report with photos the presence of numerous individuals of native ladybugs. Subsequently, professional researchers visited the ladybug population and collected individuals for captive breeding and research.

Death is another ecological phenomenon that is often difficult to study because dead animals are rarely encountered. Citizen science surveys of beaches such as the Coastal Observation and Seabird Survey Team (COASST) (Parrish et al. 2007) can provide information on timing of mortalities of oceangoing bird species. Mortality of birds around human residences also can be documented using projects such as American Bird Conservancy's PredatorWatch and the Cornell Lab of Ornithology's My Yard Counts (2006–2007), and Project FeederWatch data have been used to estimate annual national mortality as a result of window strikes (Dunn 1993).

Mortality due to rare environmental events also can be inferred by examining data from ongoing citizen science projects. Following the 2003 heat wave in Europe, researchers used data from the French Breeding Bird Survey to show that death or reproductive failure rates for different bird species were related to their predicted resilience to extreme temperatures (Jiguet et al. 2006). Because temperature anomalies are more important than absolute temperatures in influencing populations, these insights about extreme temperatures would not have been possible without data collected over a long enough time frame to include the uncommon heat wave.

Other hard-to-observe events, such as the use of calcium sources by birds, also can be documented by citizen science projects (Dhondt and Hochachka 2001). Project participants provided sources of calcium for birds and made observations of the species that used supplemental calcium during their nesting seasons. The results showed that between-species variation in the use of supplemental calcium was closely related to diet.

Challenges

Conducting ecological research using citizen science tools, whether collecting data from a new project or using data already collected, will need to overcome obstacles inherent in data gathered via volunteer schemes. Researchers need to address challenges from several angles—during project design, through project-to-participant communications, and by the use of appropriate statistical analytical methods. In general, working with citizen science data requires more sophisticated statistical and data management skills than do many other types of ecological research.

Observer Variability and Detection Probability

Participants contributing data to citizen science projects vary in age, experience, skill, training, willingness to be trained, and other attributes that influence data accuracy (Dickinson et al. 2010). This challenge can be addressed with protocols that minimize required skill levels, provide adequate training, and maximize standardization of data collection to increase consistency across observers. Protocols designed to encourage repeated observations at individual sites are valuable for two reasons. First, all observers, professional or not (Kendall et al. 1996), differ in their abilities; similarly, observation locations may differ from each other in consistent ways that cannot be readily quantified. Protocols requiring repeated observation allow these observer- and site-related differences to be accounted for during analyses, removing the intersite and interobserver biases. Second, failure to detect something indicates that either this thing (e.g., bird species) was not present or, alternatively, it was present but not detected. For example, a silent, motionless bird can be nearly impossible to detect but is nevertheless using the habitat at a location. Formal methods of analysis now exist for analyzing data from systematic, repeated counts in order to separate out probabilities of true absence from lack of detection (MacKenzie et al. 2006). At least two projects at the Cornell Lab of Ornithology (Celebrate Urban Birds and My Yard Counts) were explicitly designed to allow estimation of detectability, and other less rigorously repeated counts may also prove amenable to analysis using these same methods.

Distribution of Data

We perceive that the greatest issue with citizen science data gathering, relative to more conventional means of gathering scientific data, is the extent to which self-selection by participants affects the data-collection process. In most citizen science projects, participants have ultimate control over where they collect data, when they collect data, and the effort they expend in data collection. We concur with the conclusions of Schmeller et al. (2009) in believing that this, rather than the inherent abilities and training of participants, is why results based on citizen science data are deemed to be more biased than results based on data collected by professionals (Engel and Voshell 2002; Genet and Sargent 2003).

When participants choose the locations at which they make observations, their chosen locations may not accurately represent the local availability of habitat of different types and may not be distributed evenly across the region of interest. Even though the density of data from across the United States and Canada is roughly proportional to human population density, meaning there tend to be more data where more people live, highly urbanized sites tend to be underrepresented in the data (e.g., in eBird and Project FeederWatch). Some citizen science projects use a stratified random sampling design, for example the North American Breeding Bird Survey, the North American Amphibian Monitoring Program and several bird monitoring projects in the UK (see Chapter 12), but this is rare. With less rigorous designs, gridding the region of interest can be helpful. Most breeding bird atlas projects attempt to sample within all grids, whereas other projects specify a subset of grid squares (Kéry et al. 2005). When random spatial distribution of observations cannot be imposed via project design, analysis and interpretation need to take the nonrandom distribution of sampling points into account.

The Value of Nothing

Participants in citizen science projects can perceive their observations as unimportant, and fail to provide data, if they do not observe something "interesting." The cases with which we routinely deal involve absences: not observing one or any species, or not observing events such as presence of diseased birds. To a researcher these observations are extremely important, and our approach to this challenge has been to address it on multiple fronts. First, in project design, mechanisms can be created to turn these "nonevent" observations into something that is reportable, and their importance can be communicated. The mechanisms include providing places on data-collection forms for observers to report that they made observations but saw no birds. The House Finch Disease Survey questionnaire (a computer-scannable form) was designed so that participants had only "yes" answers (rather than yes or no answers). The first question for each day was "Have you watched your feeder?" The following questions were "Have you seen House Finches?" and "Have you seen diseased House Finches?" Therefore if participants reported that they had observed their feeder but did not report that they had seen House Finches, we knew the species was not observed. This made it possible to show that in certain parts of the range, House Finch abundance declined through winter when disease prevalence was high, while this was not the case in warmer areas (Dhondt et al. 1998).

It is possible to require participants to provide information that they might otherwise assume is not essential, such as indicating whether an electronic checklist consists of all the bird species that they identified (eBird). This latter information allows data analysts to infer counts of zero birds observed for all species present on the checklist for which no counts were reported. Second, through project implementation, communications with participants can explain and repeatedly emphasize the importance of reporting what is seen and not seen (Dhondt 1997).

Third, this issue can be dealt with in analysis as well. For example, Dhondt et al. (1998) compared conjunctivitis prevalence in House Finches reported by participants who regularly submitted data forms in the months prior to reporting conjunctivitis with data from those participants who never reported House Finches until they reported conjunctivitis. The latter participants were presumably failing to report the absences of disease and their data overestimated disease prevalence (Dhondt et al. 1998).

Handling and Analysis of Data: Researcher Training

Several other chapters in this book (Chapters 3, 7, 8, and 9) discuss methods needed to appropriately manage and analyze the types of large-scale data that can come from citizen science projects. The underlying challenge is for researchers to acquire the skills needed to understand and use the methods that are discussed in these other chapters.

Given the magnitude of data that can be collected from large-scale citizen science projects, researchers need to have strong skills in data management and manipulation. In our own work, we routinely use data files that are too large for spreadsheet programs to handle. For data such as these, researchers will need to learn to use database systems (e.g., MySQL, Microsoft Access).

We have described the uses of citizen science data in conjunction with other sources of data across large areas, such as data derived from satellite imagery. All of these data are available as geographic information system (GIS) files, and joining their data with the participant-collected data requires knowledge of the specialized software that makes use of GIS data files.

Once the data are collated and organized for analysis, the methods used for analysis can also differ from those for which ecological researchers are typically trained. Specifically, the need to use methods to account for nonindependence of data from multiple nearby locations (Chapter 7) can require relatively specialized knowledge. Regarding the exploratory data-analysis techniques described in Chapter 8, these are just starting to be used by ecologists, and training in their use would possibly require researchers to develop contacts with colleagues in the field of computer sciences rather than statistical sciences from which the "standard" analysis methods have come.

All of this means that to effectively and efficiently work with large citizen science data sets, ecologists will likely need to formally or informally develop collaborations with colleagues in disciplines with which ecologists have received little formal training or previous exposure.

🦋

Based on Thomas Kuhn's *Structure of Scientific Revolutions,* we can ask, at which phase(s) of science might citizen science have the most effect? Are there new fields that can be opened where the topics are in the "pre-science" phase, that is, prior to a central paradigm? Might citizen science move an existing area of pre-science to the "normal science" phase, that is, enlarging a central paradigm? Is it possible that citizen science will create "revolutionary science," that is, bring up enough anomalies to some current paradigm that it will create a new paradigm that subsumes the old results as well as new anomalous results into one new framework? Given the ability of citizen science to detect large-scale patterns of systems, it is possible for it to advance science in all these ways.

In this chapter, we used examples primarily from our own work to illustrate opportunities for using the methodology of citizen science. Citizen science projects have the ability to not merely contribute to understanding of the questions that initially motivated the project, but also make available a network of participants subsequently tapped to allow collection of novel data in diverse subfields in ecology, including reproductive tactics, macroecology, disease ecology, evolutionary ecology, and conservation biology.

Acknowledgments
Many of the insights in this chapter arose from the daily collaborations among staff at the Cornell Lab of Ornithology, particularly Paul Allen, Rick Bonney, and Tina Phillips. Some citizen science projects and associated research described in this chapter were funded by NSF DEB 009445 and NSF EF 062705.

7

Widening the Circle of Investigation

The Interface between Citizen Science and Landscape Ecology

BENJAMIN ZUCKERBERG AND KEVIN MCGARIGAL

Citizen science data are ideally suited to studies in landscape ecology and can support new areas of research by coupling environmental and biological data to better understand how spatial and temporal factors in the environment affect ecological processes and patterns (Turner 1989; Turner et al. 2001; Wiens 2007). In theory, landscape ecology is a highly interdisciplinary field that strives to integrate biological and quantitative approaches with research from the social sciences and natural resource management (Liu and Taylor 2002). Research topics in landscape ecology often include the study of land use and land cover change, ecological scaling, landscape pattern and ecological processes, landscape conservation, and sustainability (Wu and Hobbs 2002). In practice, however, empirical studies in landscape ecology are difficult and distinct from traditional approaches in ecology because they are typically conducted at scales that are relatively broad, either spatially or temporally or both (Brennan et al. 2002). Although past studies in ecology have provided a wealth of information by which to study and manage our natural resources, they are often limited to a particular time and place, with uncertainty as to universality of results. The emerging connection of citizen science to landscape ecology offers the opportunity to widen the circle of scientific investigation across unprecedented scales of space and time.

A major focus of landscape ecology has centered on documenting the effects of environmental patterns on the abundance and distribution of organisms (Fahrig 2005). In many cases, such research incorporates the study of habitat loss and fragmentation and its ecological consequences. The loss and fragmentation of natural habitats due to human modification is considered one of the most important environmental threats facing wildlife populations (Fahrig 2003; Lindenmayer and Fischer 2006). Scientific reviews, however, have suggested that past studies on habitat fragmentation have been too limited in scope because they focus on individual patches of habitat (patch-scale) as opposed to studying entire landscapes or mosaics of patches (landscape-scale) (Fahrig 2003; McGarigal and Cushman 2002). That is, patch-scaled studies will typically focus on a set of individual habitat patches and study the variation in a population's parameters (e.g., survivorship, abundance, occupancy) across patches in relation to patch attributes (e.g., size, shape, and spatial context). In such studies, the "landscape" is often defined as an area surrounding the focal habitat patch, that is, the spatial context of the patch. Although this approach can be useful, it fails to address fragmentation as a landscape-scale process (McGarigal and Cushman 2002).

Alternatively, landscape-scale studies seek to study pattern-process relationships across multiple landscapes supporting a range of habitat cover and pattern. What, then, is a landscape? By definition, a landscape can range from a single meter to hundreds of kilometers, but most importantly, a landscape is an area that is species specific and captures the environmental heterogeneity that is most likely influencing the study organism or biological process (Turner et al. 2001). For many wide-ranging taxa, such as birds, landscape-scaled studies are logistically challenging because ecological data must be collected across multiple, independent landscapes that may range in size from hundreds of hectares to hundreds of square kilometers. Traditionally, these data have been difficult to collect in a standardized fashion, but the emergence of citizen science and volunteer-based surveys offers exciting opportunities for advancing this line of study. The goal of this chapter is to demonstrate how the merging of landscape ecology and citizen science can be used to develop novel approaches for addressing hypotheses in landscape ecology.

Hypothesis Testing Using Citizen Science: Habitat Loss and Fragmentation

For more than thirty years, the ecological implications of habitat loss and fragmentation have received a great deal of attention in ecology (Hanski and Gilpin 1997; Harris 1984; Lindenmayer and Fischer 2006; MacArthur and Wilson 1967). Habitat *loss* involves the reduction or conversion of habitat area over time and is often, though not necessarily, concomitant with habitat *fragmentation*. Although scientists continue to debate the exact definition and implications of fragmentation (Lindenmayer and Fischer 2007), it is often taken to mean the breaking up of a habitat or ecosystem into progressively smaller and more isolated remnants (Wilcove et al. 1986).

Practically, when native habitat is lost or converted, this generally results in smaller habitat patches, larger distances between patches, and more edge between habitat types. These changes in the amount and configuration of habitat have consequences for habitat availability and may affect the ability of organisms to move and persist between habitats. More precisely, however, fragmentation is a landscape process involving the disruption of habitat continuity (i.e., physical connectedness of the landscape) *and* connectivity (i.e., the functional connectedness of the landscape as perceived by the focal organism) (McGarigal et al. 2008). It is important that, while the disruption of habitat continuity is an essential aspect of habitat fragmentation, it generally matters only if the disruption impairs habitat connectivity for a given species or population. Although habitat fragmentation can result from natural processes (e.g., wildfire, hurricanes, disease), human-induced land use changes have the ability to reshape and alter ecosystems at rates and scales that far exceed the natural range of variability for many ecosystems (Lindenmayer and Fischer 2006).

Despite past attention given to the study of habitat loss and fragmentation, many pertinent hypotheses remain to be tested regarding the cause and biological consequences of habitat loss and fragmentation. Citizen science is ideally suited to play an important role. Here are some examples:

- Most fragmentation studies tend to focus on a spatially restrictive study site (e.g., field station, county, or township). Many questions remain regarding the uniformity of fragmentation effects across entire regions. To what extent might the distribution and abundance of a species be affected differently by fragmentation in different parts of its range?
- It is generally agreed that effective conservation requires a distinction between the effects of habitat *loss* and *fragmentation,* yet there is wide disagreement in the ecological literature over which of these processes is more important and when (Fahrig 2003). Landscape-scale studies that effectively address this question require large numbers of landscapes with prespecified amounts and configurations of habitat. Is it possible to study the independent and interactive effects of habitat loss and fragmentation on populations in a real-world setting?
- Habitat loss and fragmentation may contribute to or exacerbate other environmental threats (e.g., acid rain deposition, pollution, climate change). One such threat is the potential effect of invasive species on native species and ecosystems. How might habitat fragmentation mediate the spread or persistence of invasive species, and what are the possible synergistic effects of habitat loss and invasive species on native species?

Past Uses of Citizen Science to Study the Effects of Habitat Loss and Fragmentation

There are several challenges for studying the effects of habitat loss and fragmentation on wildlife populations. In their review of experimental approaches, McGarigal and Cushman (2002) discuss six ideal characteristics that a fragmentation study should have:

1. Replication of landscapes: landscapes of similar dimensions that are drawn from the same regional area or contain accessible information as to which region(s) they originate from.
2. Biological relevance: landscape sizes that are functionally and biologically relevant to the organism or process of interest.
3. Variation in amount and configuration: a study design that allows for the assessment of the independent and interactive effects of habitat amount and configuration (i.e., how the habitat is distributed throughout the landscape).
4. Replication of landscapes: replicated and random distribution of sampling sites and corresponding landscapes.
5. Controls: adequate temporal and spatial controls so that fragmentation effects can be distinguished from natural variability in the response variable across space and over time.
6. Replication of sampling over time: a sampling period that is long enough to capture fragmentation effects, which may not be observable right away (in many cases we do not know the lag time before species and/or populations begin to respond to habitat loss).

In many settings, these characteristics are exceedingly difficult (if not impossible) to achieve. Because of logistical constraints, studies of fragmentation tend to use landscapes of limited size and are often restricted both geographically (sampling just a few sites) and temporally (sampling over just a few years). Indeed, some fragmentation studies have focused on only two landscapes: one with fragmented land cover and one with continuous land cover. The data generated from citizen science programs, however, offer an emerging alternative to traditional fragmentation studies and are uniquely compatible with the study of habitat loss and fragmentation in landscape ecology.

Citizen science projects have many attributes that make them suitable for studying the effects of fragmentation on biodiversity, and several researchers have turned to citizen science programs for this reason. As an example, the USGS North American Breeding Bird Survey (BBS) (www.pwrc. usgs.gov/BBS) is a continent-wide stratified random network of roadside bird surveys that are conducted annually by volunteers. The data collected by this monitoring program have proven extremely valuable for studying the effects of habitat loss and fragmentation on breeding birds (Boulinier et al. 1998; Donovan and Flather 2002; Flather and Sauer 1996). In one such study, Boulinier et al. (2001) acquired BBS data collected between 1975 and 1996 in Maryland, New York, and Pennsylvania. They specified a 19.7 km² landscape centered on each route, and quantified land use and land cover data within each landscape. Routes surrounded by more fragmented forest (i.e., lower mean forest patch size) supported fewer area-sensitive species and had higher year-to-year rates of local extinction than routes embedded in less fragmented landscapes. They found these patterns to be consistent across all three states, which suggested that the responses of bird communities to fragmentation are consistent across entire regions. The repeated sampling of the BBS allowed the researchers to account for the differences in detectability among species and observers. In this case, the use of hundreds of spatially referenced survey routes, an extensive sampling period (>20 years), and independent landscapes allowed for a robust testing of several hypotheses regarding the role of fragmentation in long-term population dynamics.

Another form of citizen science that has proved useful in the study of habitat fragmentation is atlas surveys. Atlases are grid-based sampling schemes that depend on volunteer participants to collect data on species' occurrences that can cover entire regions (Donald and Fuller 1998; Gibbons et al. 2007). Breeding bird atlases (BBAs) are the most common form and represent a global exercise with approximately 411 ornithological atlases being conducted in nearly fifty countries. These surveys have spatial extents ranging from 12 km² to nearly 10 million km², and sampling resolutions ranging from 0.06 km² to 14,400 km² (Gibbons et al. 2007). Atlas data are suitable for landscape-scale investigations because they offer a large number of potential sampling units (often numbering in the thousands), are nonoverlapping, and allow for an assessment of landscape structure at multiple scales. As such, these data are increasingly used for assessing species-habitat associations in landscape ecology and in the study of broad-scale population dynamics (Lemoine et al. 2007; Titeux et al. 2004; Trzcinski et al. 1999; Venier et al. 2004; Villard et al. 1999; Zuckerberg, Porter, et al. 2009). As an example, Trzcinski et al. (1999) used atlas data to test the hypothesis that the negative effects of forest fragmentation on species distributions were exacerbated in landscapes with relatively low amounts of forest cover. They treated ninety-four atlas blocks measuring 10 × 10 km as independent landscapes within a range of forest cover and analyzed the presence/absence of thirty-one forest breeding bird species in each landscape. They found that, although forest breeding birds were more likely to occupy heavily forested landscapes, forest fragmentation did not have a consistent negative effect on the distribution of forest breeding birds. By studying the distribution of birds in landscapes over a wide range of forest cover and fragmentation, they were able to conclude that avian responses to landscape pattern were generally weak and highly variable among species.

Habitat loss and fragmentation can negatively affect biodiversity by mediating the spread and colonization of invasive species. Unfortunately, studies on invasion dynamics typically lack information on the relationship between landscape characteristics and the likelihood of successful colonization by a nonnative species (With 2002). Many citizen science projects collect information on multiple species, thus these databases offer the opportunity to study complex community dynamics in relation to habitat fragmentation. For example, Bonter et al. (2010) examined colonization patterns by the Eurasian Collared-Dove (*Streptopelia decaocto*) in Florida and quantified the habitat characteristics of sites most likely to be occupied by this invasive bird species. In addition, they studied the relationship between Collared-Dove abundance and the abundance of other dove species in the study area, anticipating a negative effect on established species following the introduction of an ecologically similar competitor. By combining data collected at 444 Project FeederWatch sites with land cover data within 1 km radius surrounding each site, they found that the probability of Collared-Dove occurrence increased in landscapes characterized by low-intensity development as well as medium- to high-intensity development (Figure 7.1). Contrary to their expectations, however, the abundance of native dove species all increased, rather than decreased, with increasing Collared-Dove abundance throughout the sampling period, suggesting that invasion of the Collared-Dove has not as yet resulted in significant competition with native species.

These studies represent only a small sample of the research that has been done on habitat loss and fragmentation, but they highlight some important advantages of using citizen science. First, landscape-scaled fragmentation studies are often "data hungry" because sampling is required over a range of habitat cover to capture the underlying variability in fragmentation. Citizen science projects often collect data at thousands of sampling sites, allowing for a robust sample size, which encompasses a wide range of environmental variation. This allows researchers the flexibility to use landscapes in a random fashion or to selectively sample landscapes to reduce the expected negative relationship between habitat amount and fragmentation (allowing for the study of the independent effects of habitat amount and configuration). Second, sampling over large geographic areas allows researchers to test for the consistency of their findings across entire regions. Third, citizen science databases often (although not always) contain information on who collected the data and for how long, allowing researchers to account for differences in observer quality and sampling effort. Fourth, collecting data on multiple species of varying life history characteristics (e.g., migratory status, vagility, habitat specificity) provides scientists with the opportunity to address additional hypotheses as to which species might be more or less affected by fragmentation and why.

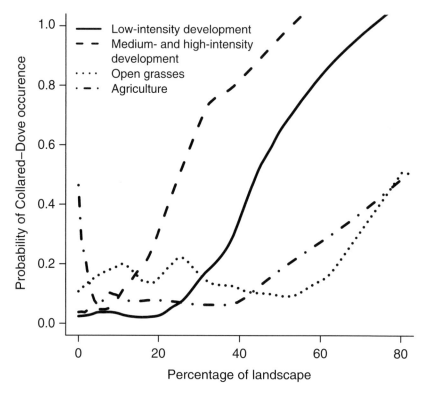

Figure 7.1. The probability of the occurrence of Collared-Doves demonstrated strong relationships with human-modified land cover types such as low-intensity and medium- to high-intensity development in Florida. These relationships provide strong evidence that, in some systems, habitat loss and fragmentation may serve as processes that mediate the spread and persistence of invasive species. Adapted from Bonter et al. 2010.

How to Use Citizen Science to Answer Questions in Landscape Ecology

Vetting the Database

Citizen science data must meet certain requirements to be useful for landscape ecology research. First, georeferenced sampling occasions or study sites are critical. Using data that are spatially explicit (i.e., sites that have an exact latitude and longitude) is necessary if the investigator plans to delineate landscapes. For example, Project FeederWatch (see Chapter 2) allows participants to use online mapping tools to identify the exact geographic location of their feeder station (Figure 7.2). These coordinates are included in the database and include information on the entry technique used by the participant for assigning coordinate information. Entry techniques can include use of the online mapping tool or a personal GPS (Global Position-

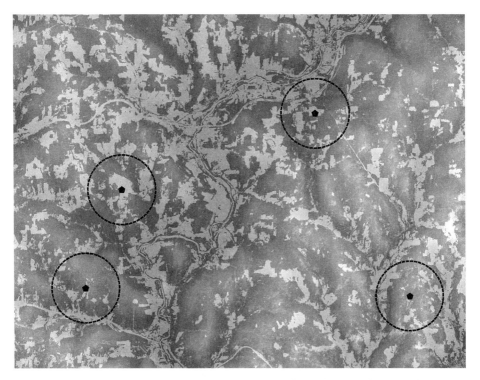

Figure 7.2. An example of four Project FeederWatch sites located in central New York. A 2-km buffer has been applied (dashed lines), and the underlying land cover represents continuous forest cover data from the 2001 National Land Cover Data program. The use of citizen science data in combination with emerging land cover data is a central component of the current and future study of habitat loss and fragmentation.

ing System). When only address information is available for a study site, geocoding can be used to associate geographic coordinates to data using street addresses or zip codes (postal codes). With geographic coordinates associated with every study site, the data can be mapped and entered into a GIS (geographic information system), and landscapes of any dimension can be delineated using buffer tools.

In addition to exact spatial coordinates, effort data are especially important for analyzing citizen science data. Strict protocols standardizing observer participation and effort are always preferred, but some citizen science programs have more flexible guidelines or benchmarks regarding participant effort. In these cases, any lack of data on observer effort makes it difficult to attribute the change in species occurrences or abundance to environmental processes as opposed to biases in sampling effort. For example, atlas surveys often have participants record the number of hours and number of observers associated with every atlas block (McGowan and

Zuckerberg 2008). In the same fashion, Project FeederWatch participants record the number of hours and number of half-day periods (over the 2-day count period) they spend watching their feeding station. These types of data are an essential characteristic of good citizen science databases because they allow the researcher to quantify the variability in sampling effort across sampling sites or remove sites that may be undersampled, or both.

Repeated observations at sampling sites have become increasingly useful thanks to advances in modeling detection probability (MacKenzie 2006). Repeated observations at the same site allow analysis of species detectability, which is the probability of observing a species given it is actually present during a sampling occasion. It is important to remember that failure to detect a species in a habitat does not mean that the species is truly absent (Kéry et al. 2008; Kéry and Schmid 2004). Cryptic or rare species that may be sampled in citizen science projects, such as amphibians, are especially prone to underdetection and false absences (Thompson 2004). Because species differ in their detectability due to various factors such as elusive behavior and their habitat associations, a more systematic approach with repeated visits is necessary to generate more meaningful analyses (MacKenzie 2006; MacKenzie et al. 2005; Mackenzie and Royle 2005). In some cases data from repeated visits will generate habitat-specific species detectability measures that can help avoid biases due to differences in detectability of the same species in different habitats. These analytical approaches have proved useful for analyzing data collected by citizen science programs such as atlas surveys (Royle et al. 2007) and BBS data (Boulinier et al. 2001).

Matching Citizen Science with Environmental Data

Citizen science data increase in value when they can be combined with environmental data of an appropriate spatial extent and resolution. Spatial extent refers to the area over which data are available and resolution is the size of a sampling unit (e.g., pixel size or minimum mapping unit). Researchers should take these characteristics into consideration when matching environmental and biological data because any species-environment associations will be constrained to the environmental variation contained within the sampling area (Zuckerberg et al. 2011). Similarly, the minimum resolution of both the environmental and biological data will affect the scope of any analysis. For example, the New York State Breeding Bird Atlas consists of over 5000 25 km^2 atlas blocks (McGowan and Corwin 2008). In this case, the extent of the biological data is constrained to the boundaries of New York State and the resolution is that of an individual atlas block. Consequently, if a researcher was interested in assessing fragmentation patterns within atlas blocks, they would need to locate land cover data of sufficient extent (i.e., statewide) and resolution (e.g., Zuckerberg and Porter

2010). Yet another important consideration is the temporal matching of an environmental and biological database, which can be a significant challenge when analyzing citizen science data collected over large geographic areas. The logistics and computational difficulties associated with classifying land cover and land use change increase with the size of the area and the number of years in the sample. Thus, for a citizen science program that has been collecting data for more than 100 years (e.g., the Christmas Bird Count) over a large geographic region, finding environmental data covering the entire temporal and spatial scale may be the limiting factor.

Since the start of the twenty-first century there has been a significant increase in the quantity, quality, and accessibility of environmental data (Gottschalk et al. 2005; Zuckerberg et al. 2011). Data on elevation, land cover, land use change, and topography are becoming increasingly available over the Internet, and quality continuously improves. Many (if not all) of these data are acquired through remote sensing technologies and represent an important tool for testing biological hypotheses with citizen science data. As an example, the USGS Land Cover Institute website (landcover.usgs.gov/index.php) is an excellent gateway for acquiring environmental data. This site contains a comprehensive list of available land cover and other environmental data sets and programs, many of which provide needed information for studies using citizen science data. One of the constraints in using citizen science data that include observations from other countries (e.g., Canada) is that environmental data are often not available or are not compatible with those from the United States. But advances in remote sensing and an ever-expanding coverage of countries are making the problem of combining environmental data from different countries less of an issue (e.g., see data sets generated by the MODIS instrument, Moderate Resolution Imaging Spectroradiometer [modis.gsfc.nasa.gov]).

Defining the Landscape and Analyzing Spatial Patterns

One of the greatest challenges confronting the use of citizen science data to investigate species-environment relationships at the landscape scale is defining the landscape in a manner that is relevant to the focal organism(s). This is often difficult given the inevitable limitations in the type and quality of biological and environmental data that can be obtained. Modern landscape ecology is largely based on the "patch mosaic" model, in which landscapes are conceptualized and analyzed as mosaics of discrete patches (Forman 1995; Turner et al. 2001), and any reading of the published landscape literature shows a near-universal adoption of this model. The strength of this model lies in the multitude of metrics available for quantifying categorical land cover patterns (Leitão 2006; Turner et al. 2001). Moreover, software tools such as FRAGSTATS, designed for analyzing patch mosaics, are readily available and easy to use (McGarigal et al. 2002).

More recently, however, McGarigal and Cushman (2005) and McGarigal et al. (2009) introduced the "landscape gradient" model as an alternative general conceptual model of landscape structure based on continuous rather than discrete environmental heterogeneity. In this model, the underlying spatial heterogeneity is represented as a continuous surface rather than a mosaic of discrete patches. The most common example of a landscape gradient model is a digital elevation surface, but there are many other possibilities with available environmental data such as percent forest cover and percent impervious surface. McGarigal et al. (2009) presented a number of surface metrics for quantifying the landscape structure of continuous surfaces, but these have yet to be implemented in readily available and user-friendly software.

❦

The early biogeographers of the nineteenth century could not have imagined the amount and breadth of environmental and ecological data being collected today. Advances in remote sensing have produced, and continue to produce, increasingly sophisticated environmental data on land cover and land use change, climate, topography, water quality, and elevation. For land cover alone, every decade since the 1980s has presented a historical milestone in remote sensing from the first 1-km Global Land Cover Characteristics database developed in the 1990s to the current systematic assessment of National Land Cover Change in the 2000s. Concurrently, citizen science is advancing and evolving at a rapid pace. As technology propels these two frontiers, the value of citizen science data will only increase, allowing researchers in landscape ecology (and other fields) to address new and old hypotheses in novel ways.

Acknowledgments
We are grateful to the thousands of citizen scientists all around the world devoted to collecting valuable scientific data, and to the countless administrators and staff who organize and oversee these programs.

8

Using Data Mining to Discover Biological Patterns in Citizen Science Observations

DANIEL FINK AND WESLEY M. HOCHACHKA

The citizen science initiatives that are the focus of this book collect large amounts of information on the occurrence and abundance of species at known and exact locations and across very large regions. By combining this information with the vast environmental data resources now available (e.g., remote sensing information on habitat, weather, climate, and soil types), citizen science initiatives can provide new insights about the factors that determine where species live, their abundance, and how environmental influences vary from region to region or through time. The opportunities promised by these data are novel landscape-level ecological insights and discoveries. The challenge is to discover the patterns that account for potentially complicated ecological relationships within very large, very noisy data sets.

In this chapter we introduce a new class of statistical tools designed to take advantage of access to massive quantities of data with a strong emphasis on pattern discovery. Similar tools are used to sort through data on e-mail users, cell phone users, and social networking sites. Developed in the machine learning and statistics communities, these data mining tools take a "data-driven" approach to analysis in which information emerges from the data during exploratory analysis. These methods have been successfully employed in many scientific disciplines that are facing similar challenges as the result of the accumulation of large quantities of data and the need for new analysis techniques to study them.

In the first part of this chapter, we describe how data mining can be used to explore and learn about the distribution of species and the insights that can be gained by applying these methods to broad-scale citizen science data. Using data from bird monitoring projects eBird (www.ebird. org) and Project FeederWatch (www.birds.cornell.edu/pfw), we demonstrate an exploratory strategy in which researchers move from investigations of general phenomena (e.g., where are birds found and when?) to investigations that identify and explain the processes responsible for the phenomena (e.g., what habitats are most important to birds in the breeding season or during migration?).

The second part of this chapter begins with a brief discussion of the distinction between exploratory and confirmatory analysis to help researchers decide when the exploratory approach we have developed is useful and appropriate. Then we discuss important practical issues that should be considered when planning an exploratory analysis with citizen science data including data requirements, the analytical control of bias, model selection, and model validation. Although our examples deal with observations of birds' distributions, the analytical processes that we discuss are generally useful in the analysis of many types of citizen science data.

Exploratory Species Distribution Modeling

Understanding where species are and when they are there is essential for many ecological studies and is critical for population monitoring, assessing conservation status, and developing management strategies. Because of the importance of this basic ecological knowledge, there has been an increasing effort to use citizen science projects to collect the broad-scale data on species occurrence and abundance. Data on species occurrence and abundance are, however, sparsely distributed in space and through time. No matter how many people participate in a project, it is impossible to monitor and record fine-scale species data with even coverage across entire continents.

To make the fullest use of broad-scale citizen science data, scientists need the ability to accurately interpolate where data were not collected (Scott 2002). By relating environmental features that are important to the species (e.g., habitat, climate, human population density) to observed occurrence or abundance data, statistical models can make predictions at unsampled locations and times. These statistical models are called species distribution models (SDMs). The success of this approach depends strongly on a scientist's ability to build a good SDM. To build a good model requires a good understanding of the ecology of a species. First, the scientist needs to know which environmental features are the most important predictors of that species' distribution to include in the model. Second, the scientist must be able to accurately specify the mathematical relationship that links the specified environmental features with the distribution. These specifications can be complex, especially for SDMs spanning large spatial or temporal scales. At large scales, many species exhibit multiple, distinct local habitat associations, all of which need to be included in the SDM.

For many species, the dominant ecological processes are simply unknown or are not understood in sufficient detail to specify as an SDM. In these situations exploratory modeling with data mining techniques offers an alternative approach for modeling broad-scale species distributions. The essential feature of these techniques is that they *automatically* adapt to patterns in data, reducing the amount and detail of information that must be supplied by the scientist. These techniques are able to screen large numbers of environmental features to identify the most important ones, and automatically determine the mathematical relationship between the important environmental features and the distribution. Practically, this means that a scientist can construct a good SDM by simply specifying a large number of *potentially* important environmental features and then letting the algorithm take care of the details. Several data mining methods have been successfully used to model species distributions (De'ath 2007; De'ath and Fabricius 2000; Phillips et al. 2006; Prasad et al. 2006) and they have been found to compare favorably with SDMs based on more traditional statistical models (Elith and Graham 2009; Elith et al. 2006; Hochachka et al. 2007).

Because data mining models have been designed to automatically discover patterns, they can be used to do more than simply "fill in" among the sparse data to produce complete species distribution maps. For example, when the dominant ecological processes for a species are unknown, an SDM can be fit and then "mined" to discover which environmental features, of those included in the analysis, are most important for predicting the distribution of that species. Once researchers know which predictors are important, they have information that can let them generate hypotheses about the actual mechanisms that affect distributions. The ability to discover and adapt to patterns in large, complex data sets makes data mining methods especially well suited for exploratory analysis and subsequent hypothesis generation (Hochachka et al. 2007; Kelling et al. 2009).

In the remainder of this section we demonstrate how data mining techniques can be used to explore occurrence and abundance data and discuss the insights that can be gained by applying pattern discovery to citizen science data from two large-scale bird monitoring projects.

eBird and FeederWatch Species Distribution Modeling Examples

Our first example is based on the analysis by Sullivan et al. (2009) in which the goal was to develop a quantitative understanding of how the Northern Cardinal (*Cardinalis cardinalis*) varies in abundance across its range. "Traveling count" observations from eBird were used. On traveling counts participants record numbers of individuals of each bird species detected visually or acoustically, the location where their search took place, the time that they initiated the search, the time spent searching, and the distance covered during the search. The traveling counts used in this example come from the conterminous United States between 2004 and 2007.

We used a data mining model to predict the observed cardinal counts. The environmental features used to predict the counts include a set of variables describing local habitat and anthropogenic activity; both of these are known to be useful for capturing abundance-habitat associations across a wide number of avian species. Habitats are classified into one of fifteen classes within the region, using data from the U.S. 2001 National Land Cover Database. To account for additional anthropogenic effects we used human population density estimates from the U.S. Census Bureau 2000 census block-level summaries. We included the time spent and the distance traveled looking for birds to account for any systematic increases in numbers of birds detected with increased search effort. Finally, we included the year, day of the year, and hour when each observation was made.

Species Distribution Maps

After the model is fit and validated (see the next section for more information on models and their validation), a map is produced to visualize what the model has learned. The map in Figure 8.1 is based on the predicted abundance at 75,000 locations selected to uniformly cover the conterminous United States. Each of these predictions is based on the habitat and human population density characteristics at that location. Thus, the model learns which habitat-abundance relationships best explain the data and then apply these same habitat-abundance relationships in new, unsampled locations.

The map in Figure 8.1 shows the expected abundance of the Northern Cardinal in the spring of 2008. More specifically, the surface shows the expected number of Northern Cardinals that would have been observed on a typical eBird traveling count beginning at six o'clock in the morning covering a 1 km transect during a 1-hour search in the spring of 2008.

The surface shows the highest predicted numbers in the southeastern United States, and the lowest predicted numbers near the northern periphery of its range in New England. The smaller Arizona population has also been identified by the model.

Identifying Important Predictors

The same predictive models that are used to produce maps such as those in Figure 8.1 will also produce insights into the basic biology of the species, because the process of making consistently accurate predictions requires identifying important predictors and quantifying their relationships with the observations. Thus, by analyzing the predictive model itself we can elucidate what it has "learned" from the data.

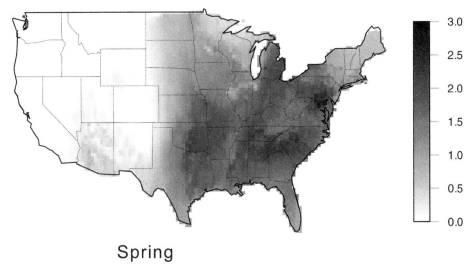

Spring

Figure 8.1. Northern Cardinal distribution in 2008. The estimated number of Northern Cardinals observed during eBird counts at six o'clock in the morning on a 1 km transect conducted over 1 hour taking into account local land cover characteristics. Darker regions indicate greater abundance.

For most explorations, the first step is to identify the most important predictors. "Variable importance statistics" analyze the structure of a model to determine how frequently a predictor is used and measure its impact on the accuracy of model predictions. These measures are useful for characterizing species habitat associations and how they can vary. For example, bird species with very broad geographical distributions must, necessarily, adapt to a variety of local habitat types. Figure 8.2 shows how habitat preferences vary for Northern Cardinals by comparing variable importance measures computed in Arizona and New York. All measures were standardized to have a maximum value of 1.0 with larger values indicating more important predictors.

In each region, the group of important predictors is relatively small, and most predictors have little importance, a typical pattern that we have observed in many analyses. The most important habitat predictor in Arizona is "shrub scrub" while the most important habitat predictors in New York are "developed, open" and "developed, low," two remote sensing categories describing parklands and suburban areas. These plots (Figure 8.2) suggest a difference between the observed habitat associations between these two regions. Such results from the analysis of citizen science data can be used to generate hypotheses to be tested with subsequent and more detailed studies.

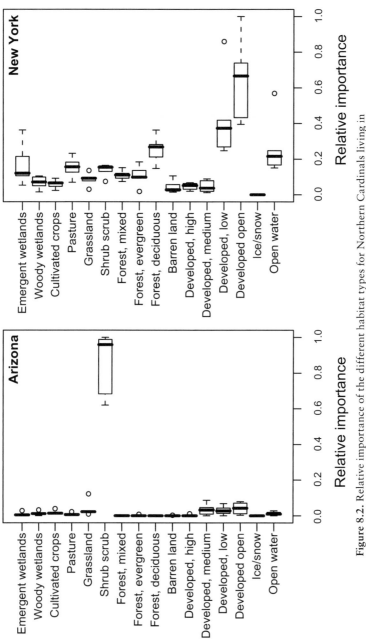

Figure 8.2. Relative importance of the different habitat types for Northern Cardinals living in the southwestern U.S. state of Arizona (left) and northeastern U.S. state of New York (right). The larger the relative importance, the more important the prediction. This plot suggests that Northern Cardinals exhibit different habitat associations in different regions of the United States.

Exploring Observational Bias

The same straightforward data-collection protocols that make it possible to engage tens of thousands of citizen science participants often produce observational data with many known and suspected biases. These biases arise when variation in the way the data are collected is confounded with occurrence and abundance patterns that we want to study.

For example, it is generally accepted that variation in species detectability is an important source of bias. Species detectability is known to vary with the amount of search effort and the expertise of the observer. It also varies with daily and seasonal changes in the behavior of a species. Some birds are easily detected during conspicuous daily activities like the "dawn chorus," whereas the same birds may become very difficult to detect when they are nesting in the summer and adopt cryptic behaviors to avoid predation. Similarly, variation in sampling intensity also affects broad-scale citizen science data; gaps in the data can be caused by differences in rates of data collection between areas populated and unpopulated by humans or across years and seasons.

These and other sources of variation in the data can contribute to sampling biases or incorrect estimates of occurrence or abundance. To make the best model, and extract as much ecological information as possible, we need to identify and account for these biases. The highly automated data mining models offer a simple way to do this when analyses include predictors that describe the bias. For example, we know that the longer participants search for birds, the more likely they are to find them. For this reason we included the number of hours spent searching for birds as a predictor in the SDM. Additionally, we included the distance traveled while searching for birds to describe variation in observation effort and we included the time of day the search began to describe diurnal variation in cardinal activity.

Techniques such as these should be very helpful in combination with more conventional analyses, such as those discussed in Chapters 6, 7, and 9, to aid in selection of data for analysis, selection of analysis techniques to reduce bias, and interpretation of results.

Isolating, quantifying, and visualizing how each of these predictor variables affects the predicted abundance can help development of strategies for controlling observational bias. Here we use a statistical tool called partial dependence functions (Friedman 2001; Hastie et al. 2001; Hooker 2007) to isolate and quantify these effects. In this instance, "partial dependence" measures how eBird abundance is predicted to change as a function of some predictor value. Figure 8.3 shows how eBird's measure of abundance is expected to change as a function of the total effort time (number of hours, left panel) and distance traveled (kilometers, right panel). This plot confirms that as participants spend more time looking for birds, they are

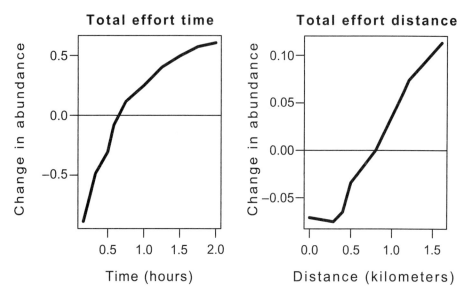

Figure 8.3. Estimated effect of effort: time (hours) spent searching for birds and distance (kilometers) traveled searching for birds. These plots show the effect of search time and search distance, after controlling for all other effects in the model.

more likely to record higher counts of Northern Cardinal. The partial dependence function also quantifies this relationship, describing exactly how each additional amount of effort will increase counts. The effect of each additional hour spent searching diminishes as the duration of the search increases. The effect of carrying out longer searches is also predicted to increase, but only after the distance traveled increases beyond a quarter kilometer. This plot provides useful information about how participants search for birds—for example, suggesting that participants survey an area with an "average" radius of about a quarter kilometer every time they stop.

Estimating Trends

Our second example is a species abundance analysis for the House Finch (*Carpodacus mexicanus*), a common backyard feeder species observed by participants in Project FeederWatch. FeederWatch participants periodically sample the birds they see at their feeders from early November through early April each year. Each sampling event consists of a 2-day count period over which they report the maximum number of individuals of each species observed at any one time. We used FeederWatch data from the eastern United States observed across 11 winter seasons (1993–1994 to 2003–2004) (see Fink and Hochachka 2009 for more information).

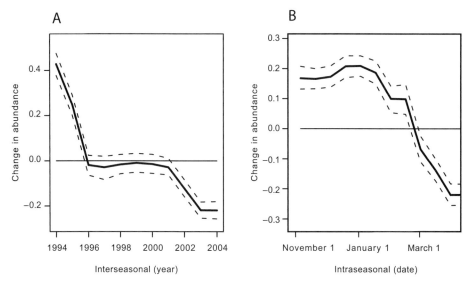

Figure 8.4. Across-year and across-season trends in reported counts of House Finches in the lower Great Lakes–Saint Lawrence plain area. These plots show how expected abundance varies across years (A) and across seasons (B) after accounting for all other predictor effects. The dashed lines are approximate 90% confidence regions.

Here we demonstrate how partial dependence can be used to estimate trends in abundance or occurrence. Figure 8.4 shows partial dependence plots of change in abundance through time for House Finches in the lower Great Lakes–Saint Lawrence plain area. These partial dependence plots show variation across years and across seasons in the numbers of House Finches reported, after accounting for all other predictor effects, with approximate 90% confidence regions. The declining trend across years (A) agrees with our expectation of a decline due to the emergence and spread of a novel bacterial pathogen, *Mycoplasma gallisepticum* (Dhondt et al. 2005). The within-year date plot (B) shows reported abundance decreasing into the spring, which may reflect the known partial winter migrations of this House Finch population, as well as seasonal variation in propensity of the birds to visit feeders.

Practical Considerations for Exploratory Analysis with Citizen Science Data

John Tukey (1977) introduced the term *exploratory analysis* to distinguish it from *confirmatory analysis*. Confirmatory techniques are largely knowledge driven, relying on the scientist's extensive prior knowledge of how to (1) identify a problem, (2) formulate specific hypotheses, and (3) generate appropriate data with which to estimate parameters of interest or confront specific predictions of the hypotheses or both. Valid statistical inference requires a tight coupling between all three of these components. As a result, confirmatory analyses tend to be tightly focused around a set of predefined relationships born of prior knowledge.

The confirmatory facet of citizen science research depends on researchers being able to generate reasonable hypotheses from theory alone, or in combination with prior exploration, to find patterns in nature and seek explanations for them. As in more traditional modes of ecological research, this discerning of patterns is a fundamental step in citizen science. The fundamental goal of exploratory analysis is to discover and describe the patterns that emerge from data, to let the data "speak for themselves." As such, exploratory analysis techniques and tools are designed to facilitate the discovery of information with a minimum of user input, in order to facilitate the creation of relevant hypotheses.

Exploratory analysis should be considered especially important when there is little prior information and no clearly developed hypothesis. For example, a landscape ecologist may want to know whether birds that migrate long distances tend to have distinct habitat associations during migration and breeding seasons. To conduct a confirmatory analysis of this hypothesis would require enough prior information to specify the seasonal habitat associations of all species being studied, a huge amount of detailed ecological information. In this situation exploratory analysis offers an efficient way to automatically learn the seasonal habitat associations. The habitat associations identified in data mining analyses can then be visualized by researchers by calculating variable importance and partial dependence values, and the patterns used to create more specific research questions (hypotheses). In general, confirmatory analyses should be used when the hypotheses under consideration have been well developed and good prior information based on related studies or theory is available. In practice, confirmatory analysis can be used when there is sufficient information to specify an appropriate parametric statistical model.

The rest of this section is devoted to the practical issues that should be considered when planning an exploratory analysis with citizen science data. Topics include data requirements, the analytical control of bias, model selection, and model validation. Our intent here is to provide more specific and detailed guidance for the analyst or practicing ecologist.

Selection of Biologically Relevant Predictors

The choice of predictors depends on the scope and specificity of the research questions, often necessitating trade-offs between predictive power and the interpretability of the predictor-response (e.g., abundance) relationships. When the goal is simply to produce the best possible SDM, we can take advantage of the data mining model's ability to extract information from large predictor sets, often increasing predictive power by adding predictors, even when the predictors are correlated. When the goal is to study the effects of a specific predictor, or a small set of predictors, the analyst must consider correlation among the predictors. When predictors are correlated, they describe the same phenomenon, thus limiting one's ability to tease apart the impact of the correlated predictors.

Predictors that describe the location and time of an observation can be used to model biological patterns in space and time, but should be included with caution because they often have strong correlations with other ecological and environmental predictors. For example, one of the main objectives of the eBird analysis was to model the Northern Cardinal distribution throughout the year. Thus it was crucial that we included the day of the year as a predictor to capture intra-annual changes in abundance patterns; this was necessary because none of the other predictors varied within the observation year. We decided not to include the latitude and longitude of an observation as predictors because the land-cover predictors already capture detailed spatial patterns.

Accounting for Observational
Bias with Citizen Science Data

As we noted above, including predictors that describe known and suspected sources of variation in the way that data are collected allows the model to distinguish between the biological and observational processes. This is important because it gives the analyst a handle with which to control or mitigate sampling bias. For example, in both the eBird and FeederWatch examples we included variables describing observer effort to control for variation in detection rates, a known bias in citizen science data.

In general, the most useful predictors of the observation process are those that vary with the collection process but do not vary with the underlying patterns of abundance or occurrence or both. In some situations a single predictor may affect both the abundance and the detection process, leading to confounding effects. For example, a species may be more abundant in its preferred habitat and simultaneously less difficult to detect in this habitat; for instance, higher densities might lead to more interactions and higher rates of singing. In general, consider including predictor information that describes the "who," "where," "when," and "how" of the data-collection events.

Exploratory Predictive Models

To this point, we have talked about machine learning analysis techniques in a general and abstract sense, but anyone wanting to use these analyses will need to select an appropriate method. Several powerful nonparametric predictive analysis methods have been developed in the fields of machine learning, data mining, and statistics. A broad overview includes neural nets (Mitchell 1997), support vector machines (SVMs) (Cristianini and Shawe-Taylor 2000), and tree-based ensemble methods (e.g., bagged decision trees, random forests, and boosted decision trees; Breiman 1996, 2001). See Hastie et al. (2001) for a good statistical overview of these methods and Hochachka et al. (2007) for pointers to available software.

Tree-based methods are especially attractive for exploratory modeling of citizen science data from wide geographical areas. Unlike many other methods, decision trees can handle a wide variety of data types automatically, both in the response and the predictors. Additionally, most decision tree implementations automatically impute missing predictor values. This high degree of automation makes them very easy to implement.

Semiparametric analysis techniques are a hybrid between the nonparametric methods noted above and the parametric statistical techniques with which most ecologists are familiar. These hybrid methods are useful when there is "partial" biological information. These methods incorporate flexible nonparametric model components to automatically discover unknown predictor effects while still allowing analysts to include specific information (i.e., in a parametric framework) that is already known about a system. Splines (Wood 2006) are probably the semiparametric technique most familiar to ecologists. In our own work with data from FeederWatch we used a parametric (random) effect in an otherwise nonparametric analysis to account for the known, consistent, site-to-site differences in reported counts of birds at feeders as part of an otherwise exploratory nonparametric analysis (Fink and Hochachka 2009). Other types of structure that can be forced on analyses include removing interactions (Sorokina et al. 2008) and forcing observations from nearby sites or close time intervals to be similar (spatio-temporal correlations) (Fink et al. 2010). Semiparametric modeling is an active research area, and new methods, useful for analyses when additional types of partial information are known and can be used, are likely to appear in the future.

Model Validation

The methods used to validate nonparametric models are also likely to be unfamiliar to most readers who have been trained in parametric statistical analysis techniques. What follows is a brief overview of the issues involved in model validation following nonparametric analysis. A model will be produced from an analysis regardless of whether the predictor variables contain any useful information. Thus model validation, the process of measuring the model's ability to make predictions over the target population, is an essential part of exploratory analysis and species distribution modeling.

The resulting measures of predictive performance are useful for model testing, diagnostics, and comparisons.

In practice, model validation is carried out by splitting the data into "training" and "testing" sets. The training set is used to develop or "fit" the model, and the test set is used to provide an independent measure of predictive performance by comparing a new set of observations to their predicted values. This sort of training-testing split can be used in the analysis of any sort of data. For example, researchers might take fifty families and subject them to a battery of tests to develop a model, then examine whether that model does a good job predicting how another, randomly selected set of families behaves when they are given the same tests. The testing set is assumed to be a representative and independent sample from the population of interest. Meeting this assumption can be very important, especially for validating species distribution analyses with large spatial extents. In general, to meet this assumption of independence the testing set observations should represent a new set of observations randomly sampled from the target population. For example, because we want to make inferences at new locations—that is, locations not included in the training set—it is important that the test set contains only data from locations that were not used for model training and that the testing locations are located throughout the study area. Similar concerns must be considered when assessing a model's ability to make forecasts or study trends. Such forecasting, however, carries the additional assumption that one can accurately describe conditions in unknown areas or at future times. An example of this additional challenge for forecasting is predicting the ability of habitat types to shift their distributions in response to climate change.

It is especially important to consider sources of sampling bias when selecting a testing set to validate citizen science models. Systematic differences between the sampled and target populations can bias validation too. For instance, consider the use of eBird data to study how forest habitat availability affects species abundance. Because eBird data tend to be collected near human population centers (i.e., areas of low forest cover), extra care must be taken to select testing set locations that adequately represent forested habitats.

Often the choice of statistic for measuring predictive accuracy is determined by the type of data: continuous versus binary (e.g., yes/no), and so forth. See Pearson 2006 for a good discussion of metrics for binary responses. Several strategies have been suggested to improve validation through test set resampling, such as cross-validation, k-fold validation, and bootstrapping. For more information on validation, see Harrell 2001 and Hastie et al. 2001.

🦋

In this chapter we have shown how model-based exploratory analyses can be used to discover and describe biological patterns and their associations with environmental features across wide geographical areas. The ability to efficiently analyze data at such large scales is important in ecology because underlying processes may change with scale such that local observations from more traditional studies do not provide the observations needed to advance research questions at broader scales (see Chapter 6). The predictive models that we describe can be used to make data-driven predictions at locations where direct observations cannot be obtained—for example, between actual observation points in areas where data are sparse. Conducting these same models also provides a framework through which we can explore and quantify predictor effects. In this chapter, we illustrated the use of variable importance rankings and partial dependence plots to identify and explore trends and examine the effect of varying effort. This sequence of model building and model exploration allows knowledge to rapidly emerge from data without requiring expert input into the modeling process.

The main analytical challenge presented by citizen science data is identifying and controlling for sampling biases, which derive from the relatively low levels of standardization and lack of control over selection of sites or effort expended by observers that is typical of many (though not all) citizen science data sets. With sufficient predictor information, explorations can be constructed to reveal and separate the effects of the observation process and the underlying ecological processes. In the examples presented here, we used effort predictors to isolate and control variation in detection rates, a common source of bias in citizen science data. By finding new model-based ways to identify and control for additional sources of bias, we can maintain the same convenient data-collection protocols while adding scientific value to the data themselves. We believe that this strategy holds promise for future ecological research using citizen science data.

Acknowledgments
Our descriptions of machine learning, exploratory analyses are based on our work to develop methods for analysis of observational data on birds for the Avian Knowledge Network (AKN). We thank our colleagues in the AKN group for the interaction and insights they have provided over the last several years. Funding for work on the AKN has come from the U.S. National Science Foundation and the Leon Levy Foundation.

9

Developing a Conservation Research Program with Citizen Science

RALPH S. HAMES, JAMES D. LOWE, AND KENNETH V. ROSENBERG

Although ecologists are increasingly aware of the role played by widespread, human-caused change in altering the population dynamics of wild animals, large-scale processes (those covering large geographic extents) remain understudied, mainly because of logistics and costs (Baillie et al. 2000; Caldow and Racey 2000). Understanding the effects of such broadscale environmental changes requires extensive data collection by volunteers to delineate patterns; elucidating underlying processes requires even more intensive study along with experimentation. In this chapter, we use the Birds in Forested Landscapes (BFL) project as an example of a research program in which a subset of volunteers was recruited to collect additional data and conduct experiments to test specific hypotheses arising from the initial observations and patterns in broad-scale citizen science data.

We have long known that the loss and fragmentation of habitat can have negative impacts on populations of some sensitive species (Andrén 1994; Robinson and Robinson 2001; Saunders et al. 1991). In addition, acid rain (Graveland 1998; Hames et al. 2006; Hames et al. 2002a), resultant depletion of soil pools of calcium (Driscoll et al. 2001; Graveland et al. 1994), and the atmospheric deposition of mercury (Rimmer et al. 2005) have been identified, along with synergies between these multiple stressors (Hames et al. 2006), as potential drivers of population declines of birds in eastern North America (Hames et al. 2002a).

BFL was originally developed at the Cornell Lab of Ornithology to address the question of how forest fragmentation affects breeding birds across North America (Hames et al. 2002b), and BFL data were later used to test specific hypotheses about the ways in which pollution affects forest birds. Here we describe how we created a conservation research program based on an intensive, fine-scale study that involved "super citizen scientists" using the BFL protocol (and extensions of it) in a small region of New York, the Catskills. These volunteers worked closely with the Cornell Lab of Ornithology (the Cornell Lab) and The Nature Conservancy (TNC) to help explore the linkage between pollutants, particularly acid rain and mercury, and declines in some species of forest birds.

Here we describe the progression of this research, our partnership with TNC, and new research using data from another long-running citizen science project, the Breeding Bird Survey (BBS), with colleagues at the Environmental Protection Agency (EPA). These current research efforts seek to understand the effects of pollution on birds over wide regions using data on the abundance of territorial Wood Thrushes (*Hylochicla mustelina*) across the eastern United States and Canada. We are using data gathered by BBS volunteers, whose efforts are administered jointly by the U.S. Geological Survey and the Canadian Wildlife Service (Sauer et al. 2004) to address hypotheses as shown in the path diagram (Figure 9.1) and to identify variation in the outcomes associated with multiple stressors, based on exploratory data analysis (see Chapter 8). Throughout, we highlight the dedication of "super citizen scientists" who allow us to address important environmen-

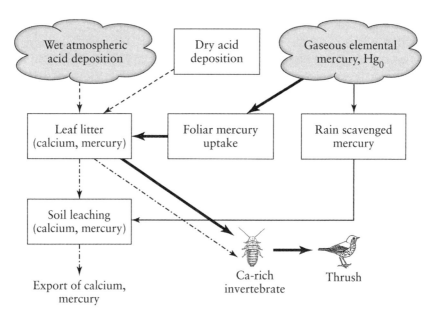

Figure 9.1. This path diagram is an explicit hypothesis on the food web pathways that we believe are important in mercury contamination of upland forest birds. Thicker lines represent what we believe are the stronger influences, for which we also have good evidence. Solid lines represent mercury (methyl and other soluble forms); the dashed lines represent acid deposition; the lines with dots and dashes are calcium.

tal questions at a hierarchy of geographic extents that is not possible with small, intensive studies carried out only by professional scientists.

Using BFL Data to Address Impacts of Large-Scale Environmental Change

Design and Implementation

BFL (Hames 2001; Hames et al. 2002b) and its precursor, Project Tanager (Rosenberg, Lowe, et al. 1999), are large, extensive, volunteer-based projects with protocols that were designed to address specific questions, such as, what are the effects of habitat fragmentation on forest-breeding songbirds and raptors? These projects incorporate more complicated and rigorous protocols than some projects mentioned in this book to ensure that detailed data on important influences on the abundance and distribution of the focal species are collected using the same methods everywhere. Standardized protocols enable comparisons between points sampled by different participants over large distances. Detailed, rigorous, and simultaneous data collection at several hierarchical geographic extents also allows us to focus on one or more factors influencing the response to fragmentation or pollution, while statistically adjusting for other influences (Hames et al. 2002a). Further, our survey protocol, including passive listening period, playback of conspecific vocalizations, and playback of small birds mobbing a predatory owl (www.birds.cornell.edu/bfl), results in a predicted detection rate for target species of 0.93, measured with the software PRESENCE (Hines, 2008). The high detection rate ensures that our BFL volunteers are unlikely to miss a breeding bird that is actually present, an important consideration in analyzing environmental effects on bird populations.

Compared with Project Tanager, which used a relatively simple protocol to study three species of tanager, BFL's more complex protocol produced improved data collection on more than forty species, with better estimates of vegetation structure, landscape configuration, and site location. Each BFL volunteer thus collected large amounts of detailed data at each site, including patch-specific data and landscape-level data. Each study site was evaluated for breeding success by the focal species using behavioral cues from breeding bird atlas codes (Smith 1990). The volunteers who persevered with this new, somewhat formidable, protocol were a self-selected group who weren't intimidated by a demanding research plan.

To date, BFL participants have returned data from approximately 3800 study sites across North America, yielding over 30,000 (species by year by location) records. Some sites were sampled in only 1 year, but many sites were resampled for up to 12 years. The improved location data available now for each study site allow for "post-processing" by matching the location to new environmental data sets that are becoming increasingly available on the Internet. We therefore have the ability to address new questions that were not contemplated when the data were first collected.

BFL Results on Fragmentation and Acid Rain

To answer specific questions about environmental change, we linked BFL data with several environmental monitoring data sets. Specifically, BFL data, originally collected for fragmentation studies (see Hames et al. 2002b), were used retrospectively to address the combined effects of pollution and habitat fragmentation, especially combined atmospheric deposition of acidifying ions and mercury (Hames et al. 2006; Hames et al. 2002a), using data from the National Atmospheric Deposition Project (NADP/NTN/MDN) (Lamb and Bowersox 2000). Additionally, we derived soil pH and other properties using the USDA Natural Resources Conservation Service's STATSGO and SSURGO soil databases (USDA/NRCS 1994). Finally, the BBS provided information on the regional, or background, abundance of the focal species (Sauer et al. 2004). Our ability to combine data from a number of sources greatly increased our understanding of broad-scale, human-caused change because it allowed us to both cross-validate and leverage BFL data with other data, including data that are hard for volunteers to collect, such as the environmental monitoring data sets mentioned above (Link and Sauer 2002; Sauer and Link 2002; Takashi 2004; Thogmartin et al. 2004).

For example, all of the above data sources were used in analyses to test the hypothesis that population declines in a forest songbird, the Wood Thrush, were related to both forest fragmentation and acid deposition (Hames et al. 2002a). We demonstrated that the probability of Wood Thrushes attempting to breed at a site declined with increasing acid deposition, after controlling for a number of other factors such as elevation, distance to patch edge, patch size, types of vegetation, and fragmentation measures. Further, we showed synergies between acid deposition and habitat fragmentation in the form of an interaction that made the negative effects of acid rain approximately 3 times greater at sites with even moderate forest fragmentation (Hames et al. 2002a). Further analyses showed negative population trends associated with low pH acidic soils as well as acid and mercury deposition (Table 9.1).

Investigating the Relationship between Acid Rain, Calcium, and Invertebrates

We then wished to further explore the mechanisms, or processes, that led to the observed patterns of declines in the Wood Thrush, using the efforts of both citizen and professional scientists. Much was already known, based

TABLE 9.1.
Performance of pollution variable cutoffs or thresholds at predicting declining population trends for the Wood Thrush within the focal 10 × 10 km GIS cell

Landscape matrix	Cutoff	e/N	Sensitivity	Specificity	False positive	False negative	% correct	Odds ratio	Kappa
Soil	≤5.6 pH	0.56	0.75	0.65	0.35	0.25	0.72	5.56	0.38
Acid rain	≤4.7 pH	0.56	0.61	0.71	0.29	0.39	0.64	3.85	0.28
Hg dep.	≥10 ng/m^2	0.56	0.66	0.40	0.60	0.34	0.58	1.33	0.06

on the work of other scientists in Europe and North America. Acid rain, for example, and leaching of soil calcium, with concomitant declines in calcium-rich invertebrates, had been linked to population declines in several species of birds, particularly in Europe (Drent and Woldendorp 1989; Graveland 1996, 1998). We knew from the literature (Bures and Weidinger 2000; Graveland 1996; Graveland and Drent 1997; Graveland et al. 1994) and our own intensive studies (Hames et al. 2006) that the number of calcium-rich invertebrates decreased as soils became more acidic and contained less calcium.

The nonbreeding diet of most insectivorous and granivorous bird species contains too little calcium to lay a clutch of eggs, thus females must consume supplemental calcium before laying (Graveland 1995; Johnson and Barclay 1996; Poulin and Brigham 2001; Tilgar et al. 2002). Parents must also provide calcium supplements to growing nestlings for development of robust skeletal systems (Bures and Weidinger 2000, 2003; Graveland 1995; Poulin and Brigham 2001). We wished to discover whether acid rain was negatively impacting Wood Thrush populations, which have been declining precipitously across their breeding range, but especially at higher elevations in the eastern part of the United States and Canada (Sauer et al. 2004), by causing decreases in the abundance of calcium-rich invertebrates.

Measuring Forest Health by Engaging Super Citizen Scientists

To investigate these complex processes, we required a simple, repeatable protocol that addressed acidification's effects on breeding birds using simple materials available to all volunteers. We therefore developed a protocol for BFL volunteers to use in sampling calcium-rich invertebrate prey available to birds foraging in the leaf litter (Hames et al. 2006).

We tested several methods to enumerate snails and other calcium-rich invertebrates in the leaf litter and chose the "cardboard trap" method (Hames et al. 2006), a simple method based on the malacological literature (Hawkins et al. 1998; McCoy 1999; Oggier et al. 1998). We first tested the methodology for one field season at forty sites in New York. At each site, we placed squares of dampened cardboard on the leaf litter, left these "traps" overnight, and returned to count the invertebrates the next morning (Hames et al. 2006).

We also needed an index of bioavailable calcium, based on invertebrate biomass, to investigate the mechanisms through which acid precipitation may negatively impact the Wood Thrush (Hames et al. 2006). We developed and tested a categorical method to estimate the calcium available to birds as invertebrate biomass in the forest leaf litter. These data were based on trapping and sorting of invertebrates into five broad taxa (isopods, millipedes, snails, slugs, and earthworms), and four 10-mm size classes (small, medium, large, and extra large). Our categorical index showed a close and highly significant agreement between estimated and actual biomass of calcium-rich invertebrates, ranging from $r^2 = 0.89$ to 0.97 (Hames et al. 2010). Further, the number and biomass of calcium-rich invertebrates were significantly related to soil properties, including acidity and calcium content (Hames et al. 2006).

By recruiting BFL participants across the East to use simple invertebrate sampling techniques, which resulted in counts and estimated biomass of calcium-rich invertebrates, we were able to test two critical predictions of the hypothesis that acid precipitation's effects on bird populations are mediated through depletion of calcium-rich invertebrates from the litter. Hames et al. (2006) ultimately demonstrated that the probability of patch occupancy by a territorial Wood Thrush declines to zero with declines in the mass of millipedes at a site; these thrushes are absent in the absence of millipedes. Such knowledge is crucial for conservation, because human-caused changes to the environment can be mitigated to some degree—but only if their effects on the biota are understood.

Collaboration with The Nature Conservancy to Measure Forest Health

The Cornell Lab has been collaborating with The Nature Conservancy (TNC) in the Catskill Mountains since 2006 on research designed to develop volunteer-based measures of forest health, to be used in their own conservation work as well as the conservation work of other organizations. Originally, TNC approached the Cornell Lab because they wanted a rigorous, citizen science–driven data-collection protocol, such as BFL's, to use in setting up a monitoring plan for the Catskill Park. The areas of interest were large contiguous forest blocks, some roadless. TNC also sought to develop a rigorous, quantitative method for measuring forest health to use in the monitoring program. They were interested in using data derived mainly from BFL sites for their "measures of success," a quantitative methodology that allows them to set numeric goals for conservation actions and assess whether the goals are reached (Tear et al. 2005).

Lab scientists, meanwhile, sought to address the effects of acid rain, soil-calcium depletion, and mercury deposition on breeding forest birds. The Catskills have some of the highest atmospheric deposition rates in the Northeast, including high rates of acidifying ion deposition (NADP NTN 2001) and high rates of mercury deposition (Miller et al. 2005). Mercury deposition, like acid precipitation in the rural East, arises mainly from the combustion of coal (Cohen et al. 2004). In its organic form, mercury is a potent neurotoxin, affecting adult birds and their offspring, causing behavioral, motor, and sensory deficits in aquatic bird species (Evers et al. 2008).

Whereas mercury in the food web is well studied for aquatic birds, little work has focused on mercury in terrestrial, upland food webs (but see also Rimmer et al. 2005). Hames et al. (unpublished data) showed low-level mercury contamination (< 1.0 ppm) linked to soil acidity to be ubiquitous in sampled upland forest birds from five regions of New York. We hypothesized that acid rain's action on soils leads to an export or leaching of calcium from the system, resulting in shortages in soil pools of calcium and decreases in the abundance of calcium-rich invertebrates. This decline, and the paucity of calcium that results, increases the uptake of trace metals in breeding birds that are likely already calcium stressed (Scheuhammer 1991, 1996). Further, some leaf-litter invertebrates are themselves at risk for mercury contamination, as they have the metabolic pathways to absorb metals, because of their high calcium demand (Beeby 1990). Additionally, some of the invertebrates in the forest leaf litter may slow their intake of food, and hence their growth, when fed contaminated leaf litter (Beeby 1990). These hypothesized effects of mercury contamination raise the specter of breeding forest birds searching more intensively to find the calcium they require to breed from invertebrates that are smaller and less abundant and that may also be contaminated with the bioavailable form of mercury, methylmercury (Figure 9.1).

Integrating Research on Mercury Deposition at BFL Study Sites in New York

To collect the data necessary to test the hypothesized effects of mercury contamination on forest birds, Lab scientists selected intensive acid rain/mercury deposition study sites arrayed along a gradient of soil pH and mercury content in the Finger Lakes region of New York. Together, TNC and the Cornell Lab also developed a monitoring project for forests in the Catskills, collecting data on soils, invertebrates, vegetation, and birds.

TNC and Lab scientists divided tasks for this new project and worked in a way that made the most of our respective institutional strengths. First, the Catskill Mountain Program of TNC recruited local volunteers, divided them into two-person teams, provided field equipment for each team, and selected the forest blocks of interest. A group of at least five BFL study sites was located in each of six large forest blocks in the Catskills, and a team of two BFL volunteers was assigned to each group of five study sites. The first site was placed in a deciduous forest block, then four subsequent sites were placed approximately 300 m apart while remaining in deciduous or mixed forest.

Because of their location, most Catskills BFL sites (which were defined as a circle with a radius of 150 m and an area of ca. 7 ha) were in 1000 ha landscapes that were more than 95% forest and showed little or no fragmentation. We were therefore able to ignore fragmentation effects and focus our investigation on acid deposition, mercury contamination, and soil properties, as well as the abundance and biomass of calcium-rich invertebrates. Lab scientists provided expertise in training volunteers on the goals and hoped-for outcomes of the project, as well as background information on the effects of forest fragmentation. We took all twenty-five Catskill Project BFL volunteers into the field to demonstrate the protocol and met with volunteers and TNC staff at the beginning, midpoint, and end of each field season to answer questions, give guidance, and show the results of analysis of data derived wholly, or in part, from their efforts.

These "super-volunteers" collected data using a "superset" of the already difficult BFL protocol; in other words, they used all of the steps in the existing protocol and gathered other data as well. In addition to hiking several kilometers, with elevation gains/losses in hundreds of meters, these super-volunteers carried out the protocol for calcium-rich invertebrates with double the number of traps. This meant carrying 4 liters of water to each site. We also worked with these volunteers to develop a protocol to assay coarse woody debris and standing dead trees, both indicators of forest health at their sites.

Through frequent interaction with Lab scientists, volunteers could see the tangible results of their work. Subjectively, we feel, based on our experience with volunteers in conservation projects, that meeting as a group with other volunteers, Lab, and TNC scientists helped keep volunteer interest and participation high. Retention was very good, with 68% returning after the first season and 44% of our initial participants still participating in the third year of data collection at our Catskill sites.

Early results of the detailed studies at sites in New York can be seen in Figure 9.2, which shows the proportion of times a Wood Thrush was detected (out of all visits) against abundance of calcium-rich invertebrates, at all five sites in each area. These results corroborate with greater precision our earlier work showing an association between acid precipitation and calcium-rich invertebrates in areas with Wood Thrush declines.

❧

While large geographic-scale citizen science is usually restricted to monitoring, we have shown that it is also a powerful vehicle for developing question-driven research projects that dissect the critical mechanisms underlying bird declines, especially when combined with small-scale intensive studies conducted by citizen and professional scientists. We believe that BFL is an example of what can be done using an approach that is focused on collection of data that address very specific questions using precise, question-driven protocols. Because the methods still involve citizen scientists scattered broadly over a landscape, as well as intensive studies addressing mechanisms with fieldwork by Lab scientists, the data lend themselves

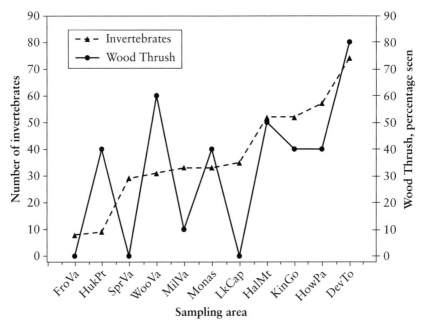

Figure 9.2. The total number of calcium-rich invertebrates and proportion of sightings of Wood Thrushes by area (five sites) in the Catskills shows an increase in thrush sightings with increases in invertebrates. Fit with a linear model, the upward trend mirrored in invertebrates and thrush sightings is marginally significant ($F_{1,9} = 5.16$, $p = 0.049$, $R^2 = 0.36$).

to the study of hierarchical processes and mechanisms underlying declines at multiple scales.

We have presented a model of a collaborative research program that is iterative, building on past results, and involves the Cornell Lab of Ornithology's Conservation Science program in studies designed to answer scientific questions at a variety of scales. Only citizen science could accomplish this research so inexpensively and expeditiously, providing scientific evidence required for environmental decision making in the face of growing human alteration of the biosphere. While large-scale monitoring is invaluable, especially in combination with newly available environmental data obtained through remote sensing, making solid biological sense of data we gathered required a network of super-volunteers willing to head out into the woods and do work equivalent to professional field biologists. We discovered a competent, dedicated set of people whose efforts have allowed us to identify mechanisms in addition to patterns. Their importance to this and other such research would be hard to overestimate.

The Conservation Science program at the Cornell Lab comprises efforts to study significant conservation issues for birds and translate science for policy and management. We have been active in a number of partnerships with governmental and nongovernmental organizations. Our efforts in these arenas have relied heavily on use of citizen science data to conduct research and to create management guidelines, species conservation plans, and high-profile reports, such as *The U.S. State of the Birds Report* (North American Bird Conservation Initiative, U.S. Committee 2009, 2010), and to inform major conservation efforts, such as Partners in Flight. This application of research to conservation policy and management is strengthened by approaches, such as we describe here, that couple large-scale citizen science with fine-scale research aimed at pinpointing causal factors in bird declines.

Acknowledgments
The authors gratefully acknowledge support from the Leon Levy Foundation for our research on forest health and bird populations. The Birds in Forested Landscapes project was developed and supported by the National Fish and Wildlife Foundation. We also are grateful for the support of the U.S. Environmental Protection Agency, a McIntyre-Stennis grant, a New York State Wildlife Grant to the Wildlife Conservation Society et al., New York State Museum's Biodiversity Research Institute, and the U.S.D.A. Forest Service. We thank our colleagues Diane E. Nacci and Carol Trocki of the EPA ORD AED/NHEERL and Daniel Fink of the Cornell Lab of Ornithology for permission to use the data in Table 9.1. We are humbled by and grateful for the dedication and hard work of thousands of BFL and BBS volunteers.

10

Citizens, Science, and Environmental Policy

A British Perspective

JEREMY J.D. GREENWOOD

Although we have not tended to adopt the term *citizen science*, collaborative work by amateurs has been an integral feature of ornithology in Britain since the early 1900s. Our citizen scientists come in two forms: members of the British Trust for Ornithology (BTO) and additional, nonmember volunteers. This is more a matter of history than of conscious decision making. In this chapter, I review the history of ornithological research as it relates to citizen science in the United Kingdom. Rather than focusing on a single research area, this chapter weaves together the historical aims of the BTO with current and historic research agendas and impacts.

The British Trust for Ornithology is the leading center of ornithological research in Britain: the work of its approximately 100 staff is extended by about 1.5 million person-hours of fieldwork by its members and other volunteers, who contribute 32% of the total volunteer input into wildlife conservation and recording in the UK (www.jncc.gov.uk/page-4253). The BTO is not a campaigning body, yet its work underpins much bird conservation in the UK and has, indeed, had major influences on government policy. The Trust's purpose and public benefit is to deliver objective information and advice, through undertaking impartial research and analysis, about birds, other species, and habitats, to advance understanding of nature. It does so by carrying out a range of activities and responsibilities:

- Sustaining long-term extensive programs and smaller-scale intensive research to study the population trends, movements, breeding, survival, ecology, and behavior of wild birds

- Encouraging, enthusing, training, and supporting volunteers to take part in scientific studies
- Bringing together professional scientists and volunteer bird-watchers in surveys of wildlife (particularly, but not exclusively, birds)
- Analyzing the data gathered through these studies and publishing the results in the primary scientific literature and via the Internet, the bird-watching and conservation press, and the media more generally

The BTO's work provides an interesting contrast to most North American citizen science projects, many of which developed in the late 20th century and with an intention to educate and engage the public in bird monitoring, rather than arising out of amateurs' concerns. The BTO has also emphasized a question-driven approach and continues to emphasize this over generalized monitoring, and its routine monitoring is not mere surveillance but involves discovering the causes of population changes and the steps that might be taken to reverse untoward changes (Greenwood and Robinson 2007).

I have chosen a broadly historical approach because knowing how the BTO has evolved helps one understand the reasons for its success—which I address in the final section. My treatment is inevitably brief: further information may be found elsewhere in print (Bibby 2003; Greenwood 2007a, 2009; Hickling 1983) and on the BTO website (www.bto.org).

Establishment

When Max Nicholson went up to Oxford as a 22-year-old undergraduate in 1926, there were few professional ornithologists in Britain and no government biological service. There were many keen amateurs but they were working in isolation. Not content to be a student, practice journalism, and write books, Nicholson founded the university Exploration Club; went on two expeditions; organized a national survey of Grey Herons (*Ardea cinerea*), which was continued annually thereafter (see Figure 10.1); set up a collaborative bird-ringing station; and laid the foundations for collaborative census work in the farmland around Oxford (Greenwood 2007b). This experience convinced him of both the value and the feasibility of an organization to draw together ornithologists for collaborative work and to provide them with direction. Gathering a team, he set up the BTO in 1933 as a society whose affairs are governed by a council elected from among the members.

The next year, the BTO mounted seven surveys—censuses of Grey Heron and Great Crested Grebe (*Podiceps cristatus*), a survey of heathland birds, and various inquiries into Swallow (*Hirundo rustica*), Short-Eared

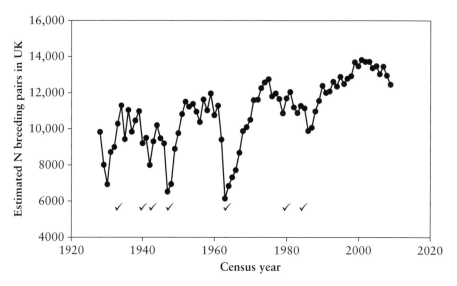

Figure 10.1. Number of breeding pairs of Grey Herons (*Ardea cinerea*) in the United Kingdom estimated from counts of nesting colonies (covering up to 75% of the nests in any 1 year) and from random sample surveys to estimate numbers in colonies missed from the main survey. The check marks show the seven coldest winters (December–February) during the period 1928–2009, from standard records for central England. Data from the British Trust for Ornithology.

Owl (*Asio flammeus*), and Fulmar (*Fulmarus glacialis*) ecology and into tameness in wild birds. All were organized by volunteers who were also responsible for analyzing and publishing the results, a pattern that was to continue for many years. In 1937 the Trust was sufficiently well established that it was able to take over the British and Irish Ringing Scheme.

Ringing

In 1907 the publisher H.F. Witherby founded the journal *British Birds* to promote collaboration between field ornithologists; 2 years later, under the auspices of the journal, he set up the ringing scheme, henceforth "the scheme." Its success was such that he soon had to appeal for donations from the ringers to cover some of his costs and eventually he started charging for the rings—a practice that continues to this day. Costs were kept down because Witherby never had to employ staff to run the scheme, much of the work after about 1930 being undertaken by Miss E.P. Leach without pay; she continued to run the scheme single-handed after the BTO took it over in 1937. The scheme was then, as now, supervised by a committee appointed by BTO Council (though with some members elected from among the ringers in recent decades); at first, every member was an amateur, but with the growth of professional ornithology in the

1950s, some professionals have been involved, all of them active ringers. The scheme happily continues to cover both Britain and Ireland, with history and considerations of effectiveness taking priority over political divisions.

Paid staff were employed to run the scheme from 1954 onward, but they were mainly involved in routine operation rather than research so most of the analyses of the data to elucidate bird migration were conducted by amateurs or by professionals working outside the Trust. Even today, after the massive and methodologically groundbreaking set of analyses published by Wernham et al. (2002), there remains a place for the amateur in the analysis of movements (e.g., Milwright 2006).

Ringing, known as "bird banding" in North America, can be used to answer questions about movement as well as survival and reproductive success of individual birds. The use of ringing to address topics other than migration is covered in various places below. The history of the BTO Ringing Scheme is described in more detail by Wernham et al. (2002) and by Greenwood (2009).

The Nest Records Scheme

In 1939, the BTO launched the Nest Records Scheme, asking observers to fill in a "card" for each nest found and record details of location, nest site, and habitat, noting date and nest contents each time a nest is visited. Over 1.25 million nest histories from 232 species have been received to date, and participants currently submit about 30,000 records annually. Several hundred papers have resulted, at first mainly on the breeding biology of single species or groups of species but latterly aimed at more generic issues, such as determining that day length rather than latitude appears to be chief determinant of clutch size in single-brooded species in Britain (Evans et al. 2009). Together, the Ringing and Nest Records schemes are the major elements in demographic monitoring (see below).

The Edward Grey Institute (EGI) and Wartime

Nicholson's vision was that the BTO should set up an institute jointly with the University of Oxford, to provide both a core of professionals who would work with the volunteers and a location for a library, a data-store, and so forth. Thus the Edward Grey Institute was set up in 1938, when it took over the census work already in progress in Oxford. During the 1939–1945 war, as part of the effort to maximize national food production, the BTO obtained government funds to work on perceived "pest" species, largely using professionals in the EGI. On the amateur side, the BTO's three long-term projects (heronries, ringing, and nest recording) were maintained, but few other surveys were undertaken.

Consolidation

Whither the BTO?

Under Nicholson's chairmanship, the Trust developed an ambitious vision after the war. Publicity and promotion were intensified, membership increased, and a paid secretary (essentially a chief executive) was taken on in 1948. That year also saw the establishment of a network of regional representatives (RRs), volunteers whose role was to promote the BTO and organize surveys at the local level based on their knowledge of their regions and its bird-watchers. Numerous surveys were undertaken, organized by volunteers, and continued for many years to come. Now numbering some 125, the RRs are a key element in the BTO's success.

In the late 1940s the EGI drew away from the BTO, which had to expand its own staff in order to develop the amateur-professional partnership that its founders had envisaged. This was probably a good thing for the Trust: had the association continued, it is likely that the EGI would have come to dominate the programs of work, with the volunteers acting merely as unpaid field-workers. My judgment, based on international comparisons, is that the fact that the members are, through the council, the ultimate governors of the Trust is a major explanation for their loyalty and thus for the BTO's success.

From Ambition to Realization

Nicholson had become a senior civil servant during the war and afterward he became a leading figure in the establishment of the welfare state. Among other things, he helped establish a government agency for nature conservation, the Nature Conservancy. After leaving the civil service, he was appointed director general of the Conservancy, which he developed into a powerful force. One of his convictions was that official nature conservation could be effective only if it recruited the voluntary conservation movement to work alongside it. Not surprisingly, the BTO was one of the voluntary bodies to whom he turned, with Conservancy funds to help develop both the Ringing and the Nest Records schemes. Staff were employed to run both schemes and there were great developments, particularly in ringing, with improved rings, improved paperwork, and improved management. Other external funding allowed a research officer to be employed in the late 1950s to develop the Trust's work on migration, in collaboration with the growing network of bird observatories (Archer et al. 2010).

Communication and feedback have always been considered key to maintaining members' enthusiasm, with both an annual report and a bulletin to report on surveys published from the start. In 1954, the bulletin, by now a multipage quarterly production, was replaced by a proper journal,

Bird Study. Bimonthly publications relating to particular schemes were added later. Both a bulletin and a journal, *Ringing & Migration,* were published for ringers. The 1950s also saw the start of conferences, an important means of keeping members in touch with the council, with staff, and with each other; national conferences of both members and ringers are supplemented with regional conferences for the many who cannot get to national events.

Long-Term Programs and Atlases

General Censuses

Although there had been targeted censuses of various single species, the BTO did not launch a general census scheme until the early 1960s, when people realized that only by long-term surveillance of large numbers of widespread species could we hope to assess the impact of modern pesticides. The Common Birds Census (CBC) involved about 200 study plots being visited by volunteers 10 times each spring, year after year; the birds and their behavior were plotted on maps, which were subsequently analyzed by staff to assess the number of territories on each plot. Changes in numbers were measured and the results reported annually, with a comprehensive set of analyses after 25 years (Marchant et al. 1990). There were occasional special analyses such as early demonstrations of the effects of Sahelian droughts (Peach et al. 1991; Winstanley et al. 1974) and of the "buffer effect" whereby populations occupy only what appear to be preferred habitats when their numbers are low, but extend into less preferred habitats as numbers build up (Williamson 1969).

The CBC was time-consuming, it was restricted to farmland and woodland habitats (though the Waterways Bird Survey covered linear waterways), and even within those it was not properly representative of the countryside because the plots were chosen by the volunteer field-workers. In the 1990s, after a period of overlap to check comparability (Freeman, Noble, et al. 2007), it was replaced by the Breeding Bird Survey (BBS), which covers all habitats through a stratified random sample of line transects, using simple distance-sampling methods. (It is thus different from the North American Breeding Bird Survey, which covers only areas adjacent to roads and uses point counts.) The fieldwork and analysis are much less onerous and over 3000 plots are now surveyed, so that region-specific and habitat-specific trends can be estimated, as well as national trends (Newson et al. 2009; Risely et al. 2009); in addition to measures of change, the recording method allows absolute population estimates to be obtained for many species (Newson et al. 2008). There is a corresponding scheme for linear waterways (WBBS).

The Wetland Birds Survey (WeBS)

By virtue of their mild winters and long coastlines, Britain and Ireland hold internationally significant populations of waterbirds in winter. After various trials extending back to the 1930s, the International Wildfowl Research Institute began national counts of wildfowl (Anatidae) in 1952–1954; the Wildfowl and Wetlands Trust (WWT) took over the organization of these in 1954. In order to cover waders (shorebirds), the BTO launched the Birds of Estuaries Enquiry in 1969–1970, soon extending it to almost all estuaries in Britain and Northern Ireland, with some coverage of non-estuarine coasts (Prater 1981). These two projects provide not only indices of national wintering populations but also information on individual sites that is of great importance for management purposes. In the "core counts" on estuaries birds are counted at high water, when they are concentrated at roosting sites; to provide information on how individual estuaries are used for feeding, a rolling program of Low Tide Counts was started in 1992/1993. The wildfowl counts and the estuaries inquiry were amalgamated into WeBS in 1993, and BTO took over the running of the whole scheme in 2004. It now covers about 2000 sites. It is poor at capturing information on species that are not concentrated on coasts or the larger inland water bodies, and efforts have been made to understand the extent to which it fails to cover these, through a survey of a stratified random sample of the wider landscape (Jackson et al. 2006).

Birds in Gardens

Gardens, or backyards, are also a special habitat, not because they contain special birds but because they are where a large proportion of the population encounters birds. The Trust started a Garden Bird Feeding Survey in 1970, to record birds that visit feeders during winter in about 250 gardens. Recognizing the widespread interest in garden birds, it set up the Garden BirdWatch in 1995 as a survey that would attract the inexperienced bird-watcher and those who had insufficient time to participate in other surveys. As in Project FeederWatch, run by the Cornell Lab and Bird Studies Canada (see Chapter 2), participants simply record the peak count of each species in their garden each week of the year (Toms and Sterry 2008). Unusually, because it costs so much to run a survey with 16,000 participants, they pay to take part. From this base, new projects on urban birds and other wildlife have been run, such as one to investigate the birds of "green spaces" in London (Chamberlain, Gough, et al. 2007).

Although garden bird projects are seen as a means of broadening the range of people who can participate in BTO work, this is not the sole aim: like all BTO work, they are designed to produce scientific data. They have been used, for example, to investigate the possible impacts of increases in Sparrowhawks (*Accipiter nisus*) on garden bird populations (Chamberlain et al. 2009), to show how variations in natural food supplies influence numbers visiting gardens (Chamberlain, Gosler, et al. 2007), and to show that daily activity in winter begins earliest in species with larger eyes and, within a species, in rural rather than urban areas (Ockendon, Davis, Miyar, et al. 2009; Ockendon, Davis, Toms, et al. 2009).

Atlases

An atlas of the distribution of birds breeding in Britain and Ireland was undertaken in 1968–1972, to complement long-term surveys (Sharrock 1976); followed by a winter atlas in 1981/1982–1983/1984 (Lack 1986); another breeding atlas in 1988–1991 (Gibbons et al. 1993); and the 2007–2011 bird atlas that covers both seasons. Atlas grid squares were 10×10 km (hectads). In the winter atlas, attempts were made to count birds; in the second breeding atlas, an element of systematic work was added, both to provide indices of abundance and to allow comparison with later atlases; the same methods were used in the 2007–2011 atlas. As well as their intrinsic interest, these atlases and the hundreds of others that have been published around the world (Gibbons et al. 2007) have been used widely both for conservation purposes and for ornithological science (Donald and Fuller 1998), especially macroecology (Blackburn and Gaston 2003).

The Ornithological Sites Register

Conservation agencies need to be aware of the most important sites for wildlife, so the BTO ran the Ornithological Sites Register during 1973–1977, to document details of the sites that participants judged to be important. In addition to the basic information, which was copied to both the statutory and the voluntary conservation bodies, the data provided the basis for a major assessment of bird habitats in Great Britain (Fuller 1982).

International Collaboration

Apart from ringing, where the BTO scheme covers Ireland as well as Britain, most BTO work is restricted to the UK. The atlases, however, have been joint enterprises with BirdWatch Ireland to get complete coverage of the two islands. In addition, the Irish Wetlands Birds Survey and Countryside Birds Survey use similar methods to WeBS and the BBS, so that we are able both to collaborate over technical developments and to produce "all-Ireland" analyses, a joint effort that makes more biogeographical sense than the separate analyses for the two parts of Ireland that are needed for administrative purposes.

British census and atlas work was done against a background of discussions at the European level, eventually leading to the setting up of the European Bird Census Council, which published an *Atlas of European Breeding Birds* (Hagemeijer and Blair 1997), drew together census data from across Europe in a Pan-European Common Bird Monitoring Scheme (www.ebcc.info/pecbm.html), and published a best-practice guide for bird monitoring (Vořišek et al. 2008).

Scientific Vision

Some of the earlier recruits to the BTO staff were, as one of them said, "men stronger in practical experience than in proven academic research" (Spencer 1983). In 1960, the leadership realized that the Trust could not only contribute large quantities of data and descriptive statistics but also apply those data to general questions in ornithology. From then on, the BTO increased its formally trained staff and the Trust developed a more modern scientific approach and vision. For example, Williamson's early demonstration that Wren (*Troglodytes troglodytes*) populations provided evidence of the "buffer effect" (described above) was followed up for several species by R.J. O'Connor, the director of the Trust from 1978 to 1987. He not only demonstrated that populations built up more in less-favored habitats as total numbers increased (and vice versa) using census data and nest records data, he also used the latter to show that reproductive success declined as numbers built up in both more- and less-favored habitats (O'Connor 1980b, 1985, 1986; O'Connor and Fuller 1985). Following the lead of the Trust's first director of science, David Snow, who used ringing and nest records data in a study of the population dynamics of the Blackbird (*Turdus merula*) (Snow 1966), O'Connor combined those data with census data and with results from a survey of garden birds. He showed that reproductive output, survival, and dispersal were density dependent and discovered an apparent buffer effect at the national level, where densities of the Great Tit (*Parus major*) were increasing in less-preferred habitats (O'Connor 1980a). He concluded, "These findings are generally consistent with or represent extensions of the results of the Oxford study of the species and suggest the technique of mass data analysis can be applied safely to less well-studied species." It was no longer possible for people to argue that intensive professional studies were the only way to address serious scientific problems.

The Trust also used its data, especially census and nest records data, to address applied problems, such as the effects of modern agriculture on birds, an important issue in a country where 70% of the land surface is farmed (Lack 1992; O'Connor 1986).

Vision into Substance

Methodological Developments

Developments in research methods have improved the effectiveness of BTO work. Pilot surveys are better designed to test methods. Fewer surveys attempt full coverage or rely on observers choosing where to work but are based on stratified random sampling. (For example, sampling may be more intense where there are more potential observers or where prior knowledge of distribution or habitat requirements indicates that a species is more likely to occur and less intensive elsewhere.) Statistical analyses are more refined and hugely more powerful, the latter enabled by increased computing power. Ready access to computers has also allowed observers to enter data online, which saves staff time and allows the results to be fed back quickly to observers and disseminated more widely through the BTO website. A standard habitat classification (Crick 1992) designed specifically for amateur ornithologists is used whenever appropriate, so that the observers get used to it and so that analysts no longer have to interpret the highly subjective habitat descriptions that observers used to submit in the past. Various training courses are provided to encourage more people to participate in the work and to increase their effectiveness when they do. The volume of work has much increased; this has involved carefully managing the demands on the observers and finding additional funding through contracts, sponsorship, and appeals. Where appropriate and after consultation with other interested parties, species other than birds have been included in some surveys.

Integrated Population Monitoring

In 1989, Baillie (1990) announced a program of Integrated Population Monitoring (IPM), designed not only to identify population changes but to assess these against established thresholds and to identify the causes of change. Core to this was using BTO census, nest records, and ringing data to build demographic models (Figure 10.2).

Routine ringing data can provide estimates of survival rates, but these estimates can be much improved if the rate of recapture of ringed birds can be increased. This was achieved through starting two special ringing projects. Constant Effort Sites, pioneered by amateurs before being taken up as a BTO project, are places where ringers make twelve visits every summer, erecting the same number of nets in the same places; because adults are usually faithful to breeding areas, plenty of between-year recaptures are obtained for survival estimates. Changes in the numbers caught provide further evidence of population changes, and the ratio of young birds to adults is an index of the year's reproductive output. Retrapping Adults for

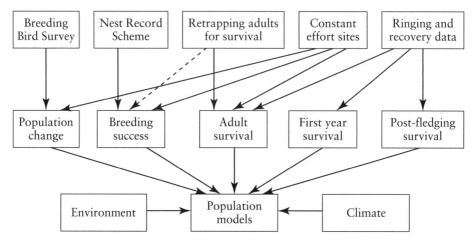

Figure 10.2. The BTO Integrated Population Monitoring program, showing how data from various annual monitoring schemes together provide estimates of demographic variables that, with environmental data, are used to build demographic models.

Survival is a scheme to encourage ringers to undertake intensive long-term studies of individual species; although individual studies may be too small to provide precise estimates of survival, it is possible to combine data from several sites to get better estimates (e.g., Robinson et al. 2008).

Once a model has been built from the data, it can be manipulated to explore which parameters best explain population changes. For example, decline in the population of the Starling (*Sturnus vulgaris*) has been accompanied by reduced first-year survival; it makes little difference to the fit of the model to the population data if other variables are held constant (rather than allowed to vary as they actually did), but if first-year survival is held constant, then the modeled population sizes depart from the observed sizes (Freeman, Robinson, et al. 2007).

Trends in demographic variables are now estimated annually for several dozen species and, using a standard set of criteria, untoward changes are flagged as "alerts" (Baillie et al. 2009). In addition, indicators of how well birds of particular habitats are doing are produced by combining data across species that are characteristic of those habitats (Figure 10.2). These are used as "Quality of Life" indicators by the government, which has made a "Public Service Agreement" to reverse the downward trend in farmland birds by 2020.

Farmland Birds

In 1990, it was generally believed that most species were flourishing, with the era of highly toxic agrochemicals safely behind us. The comprehensive review of population trends (Marchant et al. 1990) and the analyses of the data from the second breeding birds atlas (Gibbons et al. 1993) showed, however, that the majority of species characteristic of farmland had declined in the previous two decades (Fuller et al. 1995).

This prompted much research into the causes of the losses (Wilson et al. 2009). Analysis of long-term census data showed that the losses were most marked among farmland specialists (Siriwardena, Baillie, Buckland, et al. 1998) and that their timing coincided with a period of intensification and specialization of agriculture (Chamberlain et al. 2000). Ringing and nest records data showed that, for most species, it was survival rather than breeding output that had driven population changes, suggesting that the main causal factors were operating in the nonbreeding season (Siriwardena et al. 2000; Siriwardena, Baillie, and Wilson 1998). Several targeted surveys were undertaken by volunteers.

There had been a widespread switch from sowing cereals in spring to sowing them in autumn. That this might explain the decline of the Skylark (*Alauda arvensis*) was suggested by surveys, which showed that Skylarks preferred spring-sown over autumn-sown cereals (Chamberlain, Wilson, Browne, et al. 1999) and preferred weedy stubbles (associated with spring sowing) in winter (Gillings and Fuller 2001).

So great was the public interest in the losses of farmland birds that the government adopted the population of farmland birds as "Quality of Life" indicators and made a "Public Service Agreement" to reverse the downward trend in farmland birds by 2020. BTO was thus involved in identifying and testing means of doing this, based on its own and other research. For example, surveys of organic farms and nearby conventional farms indicated that while the former generally had higher bird populations, this may not have been the result of the organic prescriptions as such but of other differences, such as hedgerow structure (Chamberlain and Wilson 2000; Chamberlain, Wilson, and Fuller 1999; Fuller, Norton, et al. 2005); and surveys of fields that had been set aside from agricultural production paired with arable or grass fields showed that most birds were denser on the set-aside (Henderson et al. 2000). Other factors that were investigated included organic farming; the setting aside of fields from agricultural production; the management of arable field margins, of stubble, and of grassland; and the provision of winter food and placement of pockets of arable land within mainly pastural regions. This work contributed to various government schemes aimed at benefiting wildlife on farmland. Currently, therefore, monitoring of land that has been the subject of financial subsidies is being undertaken alongside the general monitoring program.

Woodland Birds

Many species of woodland birds have declined in Britain, though the picture is confused because others have increased. Analyses of CBC data, combined with recent resurveys of CBC plots from 15–40 years before show that declines were greatest among long-distance migrants, habitat specialists, and species that eat seeds in summer (Hewson et al. 2007; Hewson and Noble 2009). No single general explanation for the declines is obvious: it is likely that various factors have exerted a combined effect on several species (Fuller, Noble, et al. 2005). Newson et al. (2010) used BBS data to investigate whether changes in woodland bird populations were linked with the presence of Gray Squirrel (*Sciurus carolinensis*); they also assessed whether the success of nests (from the nest record scheme) was linked to the distribution of squirrels modeled from BBS data. The results indicate that Gray Squirrels are very unlikely to have driven observed general declines of woodland birds in recent years, though this cannot be ruled out for some individual species.

Climate Change

BTO research was among the first to demonstrate that climate change was affecting animals, when nest records showed that in most species with adequate data, laying dates were later in the 1960s and 1970s than in the 1940s and 1950s but had then gotten earlier in the 1980s and 1990s, tracking changes in spring temperatures in Britain (Crick and Sparks 1999). Data from the two breeding atlases showed that species that had been restricted to more southerly parts of the country extended their ranges northward in the interim (Thomas and Lennon 1999). And WeBS data showed that, as winters have warmed, wintering waders have tended not to move so far west as before, when the relatively frosty winters had made the cooler east coasts less hospitable (Austin and Rehfisch 2005).

Avian Influenza

The BTO, with its mass of researchers, is poised to respond to new threats to bird conservation. Concern over the possible spread of the H5N1 influenza virus by wild birds led to two initiatives in Britain. BTO was involved in drawing up a map showing the areas where H5N1 is most likely to enter commercial poultry flocks, based on various information including the populations of wild birds considered most likely to carry the virus; surveillance aimed at early detection of the disease can be concentrated in the areas of greatest risk (Snow et al. 2007). BTO also developed a Migration Mapping Tool (blx1.bto.org/ai-eu) that allows users to choose a region of Europe of interest and obtain information on movements of any one of twenty-one species, based on all ringing recoveries of birds moving to or from that region; maps show where birds that visit the region at some stage of their annual cycle may be found during every month of the year. From such maps, one can assess, for example, how likely it is that dead swans found in February in Scotland have come from an area where avian influenza broke out in the previous autumn.

Further Ringing-Based Research

Other analyses of ringing data have also moved beyond the traditional simple mapping of recoveries of single species in research papers. A comprehensive survey of movements, appearing in *The Migration Atlas*, was based on innovative methods for analyzing British ringing recoveries (Wernham et al. 2002). A series of papers presented information on the dispersal of birds from where they were born to where they themselves settled to breed, showing, for example, that dispersal is greater in scarcer and less widely distributed species, in wetland birds, and in migrants (Paradis et al. 1998) and that population changes are more synchronous across the country in species that disperse more widely (Paradis et al. 2002). Biometric data gathered by ringers were used to investigate (and largely confirm) that the amount of body fat that birds store is a compromise: too little fat risks starvation but too much risks being unable to escape predators (e.g., Macleod et al. 2005).

BirdTrack

In 2002–2004, BTO ran a Web-based project to document the arrival of spring migrants, replaced afterward by the year-round survey BirdTrack (www.bto.org/volunteer-surveys/birdtrack). Unlike many similar projects, while it allows the recording of individual records (the number of which received on any day reflects recording effort as much as the abundance of the species), it encourages observers to record all species they see during a bird-watching excursion. This means that the proportion of lists in which a species is recorded is a measure of its abundance, so that seasonal patterns and between-year changes can be assessed (Baillie et al. 2006).

Why Has the BTO Been Successful?

I have considered in more detail elsewhere how to successfully build citizen ornithology (Greenwood 2007a). Here I briefly list my conclusions as to why the BTO has been successful. The chief one is the close working relationship between professionals and amateurs, both well qualified but in different ways. Not only do members "own" the organization, they are involved in the development of programs—amateur ringers have, for example, been deeply involved in the computerization of the ringing scheme. Such involvement enhances their commitment to the work, as does constant feedback about the results that have been obtained. The regional representatives provide a face of the BTO at the local level, as well as being a key to getting people to participate in the work. Establishing an office in Scotland has been important in maintaining the Trust's profile there.

The fact that the BTO is responsible for almost the entire spectrum of collaborative fieldwork—censuses, nest records, ringing, special surveys—means that it has been relatively easy to integrate information from these various sources to address particular problems. The long-term continuity means that the Trust has built up a body of expertise in how best to involve large numbers of amateurs in gathering high-quality data focused on well-defined questions. And it has allowed a broad strategy to be applied to the work.

To allow the program of work to be expanded, the Trust has broadened its sources of income, particularly through contracts. The policy of accepting funding from any source except those that would compromise its scientific integrity has proved acceptable to members.

Last, the BTO has stuck to what it is good at—providing scientific information. It has not expanded its activities into areas that other bodies already occupy, such as policy advocacy. Furthermore, while it has worked closely with both official and voluntary conservation bodies, often through formal partnerships, it has maintained its independence, so that its science cannot be accused of being compromised.

Acknowledgments

I am grateful to the bird-watchers who gave me the opportunity to work with them through the BTO, to the BTO staff who have provided me with much information for this review, to Graham Appleton for critical reading of the first draft, and to Professors John Harwood and Stephen Buckland for facilities provided at the University of St. Andrews.

PART III

EDUCATIONAL, SOCIAL, AND BEHAVIORAL ASPECTS OF CITIZEN SCIENCE

The final section of this book explores the relationship between citizen science and educational, cognitive/behavioral, and social sciences research with an eye toward the potential for broad, societal impacts of citizen science programs. Educational materials are critical to ensuring proper data collection, but many citizen science projects have more significant educational goals, especially if they are funded by education grants, which are awarded based on innovative learning objectives. The potential consequences of engaging a large segment of the population in conservation research are largely unexplored. What broader impacts might we imagine for citizen science?

Future challenges for citizen science include learning about a breadth of impacts to improve project designs, and integrating knowledge of participants' abilities and learning styles to improve the usability of data for research. For example, by embedding tests of skill and knowledge into citizen science data collection schemes, it is possible to measure observer variability as a predictor of, say, what birds are seen, and also to incorporate new knowledge about what people know and how they learn into future project designs.

Citizen science can also foster inquiry. New online tools are being developed so that even large-scale citizen science can be deployed to help people ask and answer their own scientific questions (info.sciencepipes.org/about/). This will have profound consequences for engaging new audiences of individuals who may not think of themselves as scientists or as capable of engaging in scientific practice.

In Chapter 11, Jordan et al. discuss the alignment of research goals with attributes of participants, bringing together research on learning and on cognitive biases that may affect the quality of citizen science data. This exploration has much to say about future design of educational strategies for citizen science projects. Trautmann et al. (Chapter 12) focus on how citizen science can foster true inquiry in the classroom, and Purcell et al. (Chapter 13) focus on audience diversification and the removal of barriers to citizen science participation by underserved audiences.

The goals of citizen science can be expansive, quickly moving beyond education into the realms of psychology, behavior, mental health, development sociology, civic ecology or resilience theory, and theories of social learning, social networking, and collective action. The final chapters explore the contact zone between citizen science and these human dimensions impacts. In Chapter 14, Wells and Lekies explore the connection between children and nature and how this connection relates to citizen science outcomes. In Chapter 15, Triezenberg et al. integrate social networking theory with exploration of citizen science as a collective action. Next, Tidball and Krasny reach beyond outcomes for individuals to examine impacts of citizen science on community resilience within the context of civic ecology (Chapter 16). The book concludes with an afterword by the executive director of the Cornell Lab of Ornithology, John W. Fitzpatrick, whose leadership, vision, and support for harnessing the Internet as a means of delivering citizen science accounts for the diversity and continuity of projects we see at the Cornell Lab today (Fitzpatrick and Gill 2002).

11

Cognitive Considerations in the Development of Citizen Science Projects

REBECCA C. JORDAN, JOAN G. EHRENFELD,
STEVEN A. GRAY, WESLEY R. BROOKS, DAVID V. HOWE,
AND CINDY E. HMELO-SILVER

Citizen science projects explicitly link researchers and the public in scholarly research activity (e.g., Bonney, Ballard, et al. 2009; Bonney, Cooper, et al. 2009; Brewer 2002; Evans et al. 2005). The growing popularity of these partnerships raises questions regarding how to most effectively ensure data quality given the broader demands of promoting conservation and science literacy and encouraging pro-environmental behaviors. Issues related to both data quality and learning gains, which have received little attention in the citizen science literature, include cognitive processes and the potential for cognitive bias. To consider these issues, we first describe a research project with combined research and educational goals. Next we use this project to highlight the potential benefits of employing a learning sciences approach in the development of citizen science training and educational programs. Although the project is locally delivered and involves a lot of contact between project staff and participants, we end the chapter with a discussion of what our results might mean for the design of larger, Internet-based, geographically dispersed citizen science endeavors.

The Research Problem

The project, called Spotting the Weedy Invasives, tests the hypothesis that invasive plant species are more commonly found in forest patches near hiking trails than in nearby forested areas. Invasive plants often pose a serious threat to biodiversity and native ecosystems, and learning about which factors increase the likelihood of establishment of invasive species can be important to managing for conservation of native species. In New York and New Jersey forests, most invasive plants are understory herbs, shrubs, and vines, which can be difficult to locate, identify, and count (Ehrenfeld et al. 2001; Snyder and Kaufman 2004). Large-scale data-collection methods, such as aerial photography, are largely ineffective. Therefore, researchers who wish to characterize weedy plant invasions have few options beyond labor-intensive ground surveys, providing an ideal opportunity for citizen science.

Our Citizen Science Goals

We not only sought to involve volunteers in gathering data, we also had participant learning goals that included (1) gains in conceptual knowledge about invasive plant species, (2) increased science process skills with respect to collecting and analyzing large-scale data sets, and (3) enhanced appreciation for the habits of mind used by scientists. Furthermore, we wanted to document whether participants changed their behaviors regarding invasive species as measured by an increased likelihood of taking pro-environmental action, for example, by pulling invasive plants or talking with others about the problem.

Previous research has shown that people can increase their scientific understanding by engaging in citizen science initiatives (McCormick et al. 2003). Evans et al. (2005) saw increases in biological content knowledge and awareness of sense of place in most participants in the urban Neighborhood Nestwatch project (Chapter 2). Some projects suggest that involvement may even result in broader scientific literacy. For example, Trumbull et al. (2000) found that 80% of 700 participants involved in a bird feeder study engaged in scientific, inquiry-related thinking. At a minimum, by simply providing the observational tools necessary for participation, citizen science involvement usually increases awareness of scientific processes (e.g., Nerbonne and Nelson 2004; Pattengill-Semmens and Semmens 2003). In the context of environmental problems, citizen science participation has increased civic awareness and engagement as well as other environmental action-oriented behaviors (Dunlap 1992; Nerbonne and Nelson 2004; Weber 2000). For further review of public participation in research projects and the associated range of educational goals, see Krasny and Bonney 2005 and Bonney, Ballard, et al. 2009. In our study we proposed to ask whether data quality and learning impacts could be enhanced using project design strategies that integrate theories of learning and cognition.

Our Study Design and Results

Our three-year study engaged 134 volunteer hikers with little to no botanical training in collecting data about invasive plants, beginning with a daylong workshop in which they heard an invasive species lecture and then were trained to recognize the eleven to twenty-two (depending on the year) target species, carry out the data-collection protocol, and record observations. These participants, who were recruited through a regional hiking organization, were given all necessary supplies, including a pedometer to measure the distance between sample points and a simple plant press in which to preserve their collected specimens.

Data Quality

We found that volunteers were highly accurate when a subset of the data was compared to a validated data set (Jordan et al. 2011). In particular, volunteers were very good at telling whether any of the group of target species was present or not. However, at points where study plants were present, volunteers accurately identified all of the target species only 45–50% of the time, depending on the year. Volunteers (on average) were only slightly more likely to overlook a plant than to misidentify a plant, and were more likely to find plants if the plants were close to the trail and abundant. Furthermore, volunteer accuracy increased with the number of invasive plants found on the trail, which might reflect increased vigilance as more plants are spotted. When compared to botanical experts who were not trained on the project protocol, volunteers were only slightly less accurate at collecting data. These results suggest that direct experience may be more important than disciplinary training and are consistent with findings on novice citizen scientist participants in other citizen science data-validation studies (Schmeller et al. 2009).

Learning Gains

In addition to validating data with another subset of volunteers, we also assessed learning gains associated with project participation. During training, we provided volunteers with general information about plant species invasions, scientific experimental design, and taking pro-environmental action with respect to invasive plants. More specifically, after a general lecture and discussion about ecosystem ecology and invasive species, volunteers were trained in how to identify plants. Following this aspect of the project, volunteers engaged in an explicit discussion about how the data were to be analyzed. Because of the nature of the data analysis (i.e., using models to test predictions), we also made explicit the value of inference and model-based reasoning in science. Working through examples, volunteers practiced making inferences using data models. To assess learning gains, we asked questions gauging content knowledge about invasive species. We also asked questions aimed at understanding participants' behavior and actions following Marcinkowski 1993. Influenced by Lederman et al. 2002, we examined participant understanding of the nature of science, which we define as the entire scientific endeavor including the practices of generating scientific knowledge. Finally, we asked questions, modified from Etheredge and Rudnitsky 2003, that tested knowledge of the scientific process.

Our data support the notion that participating in a citizen science project with an embedded educational program can result in increased and lasting (i.e., up to 6 months) understanding of content knowledge and an increased awareness of the environmental issue under consideration. Participation, however, did not increase understanding of the nature of science and process of science. Our data also indicate that participation led to little involvement in the issue of invasive plants after the study was over (i.e., half our group reported sharing what they learned with others, but only 5–10% reported taking direct action). We did learn that some of our volunteers brought with them a broad understanding of science and skills that enable systematic reasoning (e.g., the concepts of correlation vs. causation and methods of controlling variables). We believe that this reduced the likelihood that we would find learning impacts after (compared with before) participating in the project.

Guiding Future Design: Are We Teaching the Right Skills?

We have described our native plant study because we believe that it helps illustrate how thinking about cognitive biases and learning research can help citizen science program developers anticipate issues that may arise with data collection. Our initial findings showed (1) participants were accurate at telling whether or not target plants were present, although this task may have been quite easy because invasives were absent most of the time; when plants were present, (2) participants' accuracy was greatest when targeted species were abundant; and (3) while content knowledge gains were clear, improved understanding of the nature of science and increases in direct behavioral action were not noted. Based on these findings, we consulted a cognitive scientist and the science learning literature to improve our educational and training materials.

We sought to improve volunteer accuracy and to increase the probability of achieving learning impacts among self-selected volunteers. In addition, we sought to explore the possibility that volunteers could apply what they learned from our projects in new contexts (i.e., knowledge transfer). To make these improvements, we first divided the experimental data-collection task into several components as shown in Table 11.1. Regarding data accuracy, we thought about volunteer action as a three-step process: (1) learn to identify and recall species of interest, (2) detect the species of interest against the backdrop of the field environment, and (3) appropriately follow the data-collection protocol. In addition to accurately completing these tasks, we refined our learning goals to determine whether volunteers would not only (1) retain content and (2) develop habits of mind with respect to scientific thinking, but also (3) be motivated to change behavior with respect to the environmental issue at hand.

Improving Volunteer Accuracy

One benefit of integrating learning and cognition research is that it enables program developers to strive to figure out the best way for novices to learn a new task or concept. We have an easy metric with which to evalu-

TABLE 11.1.
Goals and associated tasks

Goal	Generalized task	Specific to the protocol
Identify species	Object identification and memorization	Allow learners to first sort specimens, using compare and contrast. Trainers guide learners based on these contrasts, first working with most obvious differences and then moving to more subtle differences.
Detect species against field backdrop	Discrimination and automaticity	Allow learners to practice identification of specimens first in environments where species are obvious and then moving to more subtle situations. Allow learners to develop a gestalt.
Follow protocol/acquire process skills	Accuracy and automaticity	Allow learners ample practice with feedback from trainers.
Gain conceptual knowledge	Memorization and recall	Provide learners with an informative lecture embedded with discussion and reflection.
Appreciate nature of science	Epistemological understanding	Allow time for discussion, practice, and mistakes related to conducting an investigation. This includes involvement and growing autonomy with data collection, analysis, and interpretation.
Modify behavior	Motivation, activation, breaking of habit, and learning transfer	Show learners what steps should be followed to take action initially. Next, allow learners to take control over action, with guided learning support. Have learners not only address specific strategies but provide explanations as to why the strategy is most effective albeit not always optimal.

Note: We associated project goals with generalized learning tasks. We then designed a protocol to help volunteers achieve competence in identification.

ate task competency: the ratios of true plant identifications (successes) to false identifications (failures) for volunteers engaged in our study. Rather than approaching the training from the perspective of the experts (i.e., project directors), we can focus on the task at hand, which is how people learn to differentiate objects. This involves moving away from experts' tools (e.g., taxonomic thinking and use of keys) to focus more broadly on completing only the task needed to collect accurate data (Table 11.1). If it works well, this is a general principle that could be adopted for any citizen science project.

Consistent with stimulus-detection theory (i.e., detecting signals in the face of environmental background noise; see Tanner and Swets 1954), we are seeking to train our observers to efficiently spot the plants in question against the wooded backdrop. Given the features of object recognition, several theories of perceptual learning have been generated. Goldstone (1998) describes four mechanisms: (1) differentiation involves learning to separate once indistinguishable elements, (2) attention weighting results in increased attention to diagnostic features, (3) imprinting results in developing specialized receptors for the object, and (4) unitization results in developing an image for single construed units or objects. As natural object recognition is a result of perceptual differentiation versus differentiation of concepts as with memory for words and facts (Gerlach et al. 1999), we hypothesize that training volunteers to differentiate elements of the objects and further encouraging attention to particularly distinguishable features will enable our volunteers to search quickly for an object against the backdrop of other objects. These features, however, could be derived from what learners deem to be relevant and need not bear any resemblance to features that experts think they themselves use.

With respect to the plants in question, understanding of these mechanisms might be used to develop training modalities that will help volunteers identify species correctly. Trainers, for example, could start by having volunteers describe features they deem most visible in particular plants. After critical features of a plant are identified, trainers could highlight these parts of the plants and encourage the volunteers to draw relations between these features. This could be done by encouraging volunteers to separate plant cuttings into piles in which volunteers are asked not only to point out which plants are similar and which are different but to discuss the features that most define these differences. Consistent with Schwartz et al. (2005), trainers could begin with maximally different cases, and then move toward more subtle contrasts. Comparisons could be both physical and functional; for example, a volunteer might note that physically, a cutting with no terminal leaf resembles the ears of an animal such as a rabbit; or functionally, a volunteer might note that any plant attached to another is considered a vine. Further, to promote automaticity (i.e., the ability to complete a task with low-level processing), one could train the volunteer to develop a search image in which certain features or a unitized whole are imprinted. This "gestalt" results from rapid cognition, which involves instantaneous integration of complex information that may even be subconscious. Moving volunteers through this entire process, from coarse to fine discrimination, may be necessary to achieve a high level of accuracy.

Similar methodologies could also be used over the Web. For example, with birds, volunteers could dissect and practice the diagnostic features in the *Inside Birding* films on the Cornell Lab of Ornithology's All about Birds website (www.allaboutbirds.org/NetCommunity/Page.aspx?pid=1270). These films teach people how to sort birds into categories based on their shape, size, location, and behavior and allow for self-testing to improve accuracy.

As our learning goals include the methods and process of science, it is important to embed reflection throughout the process. Often training programs seek to start with foundations that may not be directly relevant to the task. For example, in our project we began with an extensive lecture and discussion series on ecosystem processes and ecological impacts of plant invasions. We then followed with the training protocol. Many volunteers reported that certain ideas did not make sense until after they had been in the field. Might they have benefited from hearing this foundational knowledge after they had experience with the context? Additionally, could ideas presented early in the training program be a diversion to volunteers who are already trying to learn a novel protocol?

Cognitive Biases

We suggest that pilot testing approaches based on a variety of cognitive models can greatly enhance the quality of training programs for citizen science, whether those programs are delivered online or in person. Presumably in any citizen science project, volunteers will be expected to arrive with or develop some element of data-collection skill. This skill could be the ability to identify species, recognize changes in life history, distinguish physical aspects of the landscape, or read a data-collection instrument. Next the volunteers will likely need to incorporate this skill by following a data-collection protocol; some projects might require the volunteers to engage in more varied aspects of scientific study such as experimental design or data interpretation and communication. Developing these skills and adhering to a protocol in the face of authentic data collection requires fundamental knowledge of the protocol and why it is important, which requires developing cognition in terms of sensory perception, interpretation, memory, and in many cases abstraction. To aid in the efficiency of general thinking processes, decisions are often made in the absence of complete information, leading to cognitive biases.

Here we consider cognitive biases to be impartial inclinations in thinking processes, which include sensory awareness, perception, reasoning, and memory. These biases can often be typified among people, and a number of biases have been formally described in the primary literature (see Haselton and Nettle 2006). Understanding these biases, a few of which are described below, especially as they pertain to novice learners, can be important in the design of citizen science programs.

Although it is difficult to generalize biases across the diversity of citizen science project types, certain bias categories are worth noting. Attentional biases are those that can influence what volunteers might be looking at. For example, people take longer to name colors when the words for those colors are spelled out in colored letters that do not match the name (i.e., the Stroop effect). In other instances, individuals have shown a bias whereby they tend to focus on potentially negative versus positive stimuli (Peeters and Czapinski 1990). Such attentional biases, however, can be attenuated in the presence of pertinent information. Negative attention biases, for example, may become less negative when positive information is made available (Smith et al. 2006). This can be especially important when considering statements made by volunteer trainers.

Cognitive biases in memory also exist. The context effect, whereby memories formed in one context (e.g., the training environment) can be difficult to retrieve in another context (e.g., the field environment), is particularly problematic for citizen science. Additionally, memories can change over time with certain details becoming sharper and others less so (e.g., leveling and sharpening; Wulf 1922 referenced in Koriat et al. 2000). This memory-related bias suggests that projects requiring repeated data collection might reinforce learning, avoid these biases, and produce more accurate data than brief "snapshot" projects do. Some cognitive biases, however, can be used to guide training. For example, the spacing effect occurs when ideas presented over time yield better memory outcomes than those presented more intensively within a short period of time (Challis 1993). To take advantage of this effect, projects might want to train in stages or supplement main training events, such as training workshops, with continuous interaction.

Other types of biases include those involved in making decisions about information. Framing effects are a good example of this. The way that a problem is framed can bias the decision made (Tversky and Kahneman 1981). This framing effect can be important when considering the way in which the data-collection procedure is presented. When some options are presented as being more certain than others, for example, individuals might favor those options regardless of the other pertinent information. Anchoring is another phenomenon that can occur when individuals give undue weight to certain features that skew their decisions accordingly (Kahneman et al. 2006). In a similar vein, confirmation bias describes the situation in which individuals favor information that is most consistent with what they already know (Nickerson 1998).

Certainly, the act of testing ideas in science has lent itself to experimental bias. Perhaps most notable is expectancy bias, where an experimenter is likely to find outcomes consistent with those expected (Sackett 1979). Most important, however, is making volunteers aware of the potential for bias and encouraging these volunteers to make judgments objectively using maximal amounts of relevant information. By providing training on objectivity, program developers may be able to ameliorate the effects of bias.

Because of the varied nature of many citizen science tasks, it is impractical to assume that all potential biases can be anticipated by consulting the literature alone. The best way to detect biases is to plan small pilot studies of volunteers engaging in the project tasks. Here project designers can determine likely sources of error, which may be task specific. Additionally, biases can be uncovered by encouraging volunteers to engage in a "think aloud" protocol describing their thoughts as they engage in particular tasks. Discovering biases and designing instructional materials to help participants become aware of them teaches science literacy while helping to reduce sampling bias.

Scientific Literacy and Behavioral Change

Scientific literacy, at its most basic level, is the development of habits of mind that foster systematic reasoning about scientific problems and issues. The route to such habits of mind is often circuitous, but to the extent that direct experience plays a role, it is worthwhile to think about how education can be embedded within citizen science projects to achieve these goals.

The appreciable gains we saw in basic conceptual knowledge of invasive species were expected based on past studies of undergraduate education in which the lecture is a primary mode of instruction (e.g., Saunders and Dickinson 1979). We do not know how important content knowledge is to correct reasoning about invasive plant issues, but evidence suggests that increased content knowledge may reduce flaws in systematic reasoning (Sadler and Zeidler 2005). It should come as no surprise that awareness and reflection are critical for the development of reasoning skills (Bransford et al. 1999; White and Frederiksen 1998). If we accept the view that major differences between expert and novice knowledge reflect differences in level of refinement and reorganization of ideas, and move beyond the simpler notion that volunteers hold incorrect ideas (Bransford et al. 1999; Hmelo-Silver and Pfeffer 2004), then we will design citizen science programs to promote awareness and reflection with the expectation that these reasoning skills will improve with practice. Given that learning is highly social and context dependent (Brown and Collins 1989), we could also provide meaningful scenarios within which participants are able to apply their skills. For example, rather than having volunteers perform a small aspect of the plant identification task, they could be given the opportunity to practice and refine their skills through mock simulations of the data-collection task delivered over the Web, especially if the Web application allowed volunteers to assess their own performance relative to others', and included exercises that overtly demonstrated complex issues, such as sampling bias. This gamelike approach could have positive impacts on both education and data quality.

Citizen science projects, by virtue of engaging volunteers in authentic science, pose an excellent opportunity to broaden their understanding of the nature of science. The experience of collecting and analyzing data in the context of the scientific enterprise can increase participant understanding not only about the nature of scientific inquiry but also about the general practice of scientists in terms of interpretation and communication. Teaching inquiry skills, however, can be difficult (Chapter 12). Specific training in scientific inquiry (e.g., through a targeted approach, as suggested by Etheredge and Rudnitsky [2003]), paired with actual examples of how scientific inquiry is used in other contexts beyond the questions at hand, would likely facilitate retention and transfer of these skills (Bransford et al. 1999). Offering opportunities for long-term participation with graduated levels of involvement might account for some of the successes in programs such as those offered through the Cornell Laboratory of Ornithology. It would be interesting to know whether people move from entry-level projects to more challenging projects as they gain knowledge and skills. Finally, practitioners might consider developing opportunities for design and analysis of scientific data, particularly within a social context, where volunteers can practice and engage in dialogue about pattern interpretation, learning from their own errors and those of others. This idea challenges Web-based projects to explore interactive learning technologies, including use of machine learning algorithms that map the knowledge gain of learning communities. Web-based programs, for example, could be used to track and organize discussion threads. Text analysis software could be used to categorize key terms, allowing program leaders to monitor large streams of discourse and address critical areas where knowledge gains in the community could be helpful both to the participants and to the project.

If participants are to take environmental action, which is a goal of many current citizen science projects, it seems that knowledge and practice with inquiry are not sufficient to facilitate environmental decision making (Arvai et al. 2004). Participants also need to be motivated and feel that their actions result in change. For example, according to Stern (2000), individually held values lead to beliefs that create personal norms, which are then translated into behavior. Action, however, also requires that participants overcome actual or perceived constraints (Tanner 1999), habits (Biel 2003), and perception of risks (Slovic 1987). Cognitive and emotional biases may keep people from thinking about large and overwhelming problems (Kollmuss and Agyeman 2002). This can be ameliorated by moving beyond the nature of a problem and focusing on personal agency (i.e., sense of controlling one's own actions) and self-efficacy (i.e., sense that one's actions are effective), as well as pointing out how people can be part of the solution. Projects that are large in scope, such as species invasions, climate change, or urbanization, may lead to a paralysis of action or cause people to engage in antienvironmental behaviors that provide psychological safety (Dickinson 2009). If participants feel their actions do not matter, why then would they take any? Perhaps project designs that integrate smaller-scale successes will help promote a sense of efficacy. Targeted training in decision making is also likely necessary (Edelson et al. 2006). In the example of invasive species management, this training could focus on how management decisions are made and how consensus among stakeholders is achieved.

Cognitive Aspects of Preparing Participants for Transfer of Learning

Before engaging in a complex training program, one could ask, to what extent does the project require transfer of skills among contexts? The answer will likely depend on the nature of the project and the likelihood that volunteers will encounter unforeseen issues surrounding data collection and protocol interpretation. Learning transfer occurs when an individual applies what they have learned in one setting to another (Gick and Holyoak 1980). Training for knowledge transfer often requires allowing opportunities for learners to reflect on what they have learned, practice (which often means making errors), and establishing not only a conceptual understanding of the task at hand but also a way of thinking that allows the individual to continually learn from new information (Bransford et al. 1999). Clearly, teaching for transfer is iterative and will take more than just the time required to train volunteers to complete circumscribed data-collection tasks. Doing so, however, may be worth the time if the task in question is likely to present the volunteer with variable situations in which volunteers' reasoning abilities will enhance the quality of data collected.

❧

Extensive protocol validation and educational assessment early in the process is extremely valuable in helping improve data quality for citizen science projects. For example, our ability to inspect the learning gains as well as volunteer successes and failures in invasive plant identification was extremely valuable in developing a new training protocol, which will likely increase the accuracy of data collection and improve learning outcomes. By consulting with a cognitive scientist and exploring the learning sciences literature, we have used this example to illustrate how future projects, including those delivered via the Web, may benefit by considering cognitive processes during protocol development and testing. This would allow for repeated improvements to learning materials and engagement strategies to match data-collection protocols with research goals and to avoid error and bias. New technologies may aid protocol development for projects working at large geographic scales. For example, machine learning algorithms may assess observer skill and bias in an automated fashion, while helping discover how people learn specific tasks and generating new "smart" learning environments that rapidly adjust to learning styles. The field will benefit from further consideration of cognitive biases and how people learn by testing empirically and theoretically informed assumptions about learning during front-end evaluation of instructional materials for citizen science projects.

Acknowledgments
We thank David Mellor, Kristen Ross, Edward Goodell and the New York New Jersey Trail Conference, Les Mehrhoff and Invasive Plant Atlas of New England, and Edwin McGowan of PIPC. All work was conducted in accordance with Rutgers Institutional Review Board policy. We were funded by USDA NRI award number 05–02221. Special thanks go to Joan G. Ehrenfeld, a coauthor of this chapter, who, sadly, died during the production of this book. She was a special colleague and friend.

12

Who Poses the Question?

Using Citizen Science to Help K–12 Teachers Meet the Mandate for Inquiry

NANCY M. TRAUTMANN, JENNIFER L. SHIRK, JENNIFER FEE, AND MARIANNE E. KRASNY

Use of citizen science in school settings offers potential rewards for everybody involved—scientists, students, and teachers. For scientists, inviting student participation in research offers opportunities for outreach that can build scientific literacy and environmental stewardship values in the next generation (Berkowitz et al. 2005; Brewer 2006; Krasny and Bonney 2005). For students, citizen science provides meaningful connections to the natural world through observation, data collection, and in some cases independent investigation. This engagement can be highly motivational for students, especially when coupled with the chance to present work to an audience beyond the teacher and classmates (Evans et al. 2001; Fougere 1998; Harnik and Ross 2003a; Lawless and Rock 1998; Means 1998; Moss et al. 1998; Tudor and Dvornich 2001). For teachers, citizen science offers a way to motivate and inspire students through participation in investigations that are relevant and real.

For example, the following scenario represents the experiences of a 5th-grade teacher ("Mrs. Finch") seeking to build citizen science into her classroom practice. Mrs. Finch has been interested in birds and birding for many years. Over the past 3 years, she has more fully integrated birds and citizen science into her teaching:

Year 1. As part of her life sciences unit, Mrs. Finch needs to teach about biodiversity and scientific observation. This year, she takes her students into the schoolyard for 2 consecutive days. On the first day, students sketch a wild bird and note how it uses the schoolyard habitat. On the second day, the students observe another bird and note its behaviors. Later, the students compare birds and talk about the similarities and differences they observed.

Year 2. Mrs. Finch again plans to teach a unit on birds, but this year she wants to try some new ideas she developed while participating at home in the eBird citizen science project. She starts by sharing with her students her enthusiasm for collecting and sharing bird data, and invites the class to give it a try. Working together as a group, the class learns to identify birds using online and print field guides. They practice their bird identification skills outside in conjunction with the sketching and observation activities carried out in Year 1. Finally, the class conducts 5 counts, 1 every other day. Mrs. Finch enters the data into eBird using her home computer, but uses the count data in calculations during math class.

Year 3. Mrs. Finch builds on her Year 2 bird science and math unit. This year, the class counts birds every Friday during the fall, and students post the data using eBird's online forms. The students wonder how their observations compare with data collected in other parts of the state, so they query the eBird database to find out. Discovering that one of the birds they have viewed appears to be at the northern edge of its range, they decide to continue their observations into the winter months to see whether this species will stay around or perhaps migrate south until spring. Mrs. Finch uses the accumulating data to teach basic math calculations and graphing skills. Viewing their graphs, the students interpret their findings in terms of seasonal trends of selected bird populations, and they write reports for potential publication in the Cornell Lab of Ornithology's *Classroom Birdscope* magazine.

Depending on the degree to which students design their own investigations and analyze and interpret the results, their involvement in citizen science may build science process understandings and analytical reasoning skills. In this chapter, we explore how citizen science can support student investigation, a key feature of classroom science but one that many teachers find hard to achieve.

From Surface Knowledge to Deeper Understandings and Skills

The need to build student ability to think scientifically, moving beyond memorization to critical analysis of evidence, constitutes the cornerstone of both current and long-standing efforts to reform science education (National Research Council 1996, 2000; National Science Foundation 2009). The National Science Education Standards, for example, state that the central strategy for teaching science should be "inquiry into authentic questions generated from student experiences" (National Research Council 1996:31) and define inquiry in terms of students asking questions, planning and conducting investigations, using appropriate tools and techniques, thinking critically and logically about the relationships between evidence and explanations, constructing and analyzing alternative explanations, and communicating scientific arguments (ibid., 175). More recently, the National Research Council's Committee on K–8 Science Learning defined science proficiency in terms of knowledge and reasoning skills needed by all educated citizens: knowing, using, and interpreting scientific explanations of the natural world; generating and evaluating scientific evidence and explanations; understanding the nature and development of scientific knowledge; and participating productively in scientific practices and discourse (National Research Council, 2007:36).

These proficiency goals cannot be achieved solely through lectures, textbooks, and verification-style labs. Rather, they require students to gain experience in designing, conducting, and communicating about investigations. A randomized control study that examined the effectiveness of inquiry-based versus traditional instruction on fifty-eight teens who received 14 hours of instruction over 2 weeks revealed that those in the inquiry group reached significantly higher levels of achievement across a range of learning goals that included knowledge, reasoning, and argumentation (Wilson et al. 2010). Moreover, the group that had been randomly assigned to receive traditional science instruction showed a detectable achievement gap by race, whereas the group exposed to inquiry-based instruction did not.

Approaches to student inquiry range from activities that are closely structured by the teacher to open-ended investigations in which students are responsible for shaping their own questions, procedures, and analysis techniques (Figure 12.1). Depending on desired learning objectives, teachers may choose to use a mixture of different levels of inquiry in their teaching. Highly structured activities focus attention on particular concepts or processes, whereas open-ended investigations provide greater opportunities for development of scientific reasoning and process skills. Although open-ended inquiry may be challenging in classroom settings, small steps toward inquiry can lead to larger ones as teachers gain confidence and realize the impacts on student motivation and in-depth learning.

The opportunity to pose their own questions can bring personal relevance and meaning to student investigations, inspiring interest and motivation to learn (Shirk 2004; van Zee et al. 2001; Windschitl and Buttemer 2000). Asking and reflecting on "testable" questions can foster a deep understanding of scientific concepts, develop understanding of the ways in which scientists study the natural world, and support skill development such as making observations and inferences, weighing competing explanations, and drawing evidence-based conclusions (Drayton and Falk 2006; van Zee 2000; Windschitl and Buttemer 2000). The final step of communicating and justifying proposed explanations can be highly motivational, inspiring students to achieve deeper levels of learning (Trautmann 2009).

Despite the benefits that students gain through designing their own investigations, transitioning from traditional classroom learning to an envi-

Level of inquiry	Posing a question	Designing the procedure	Analysing the results
Confirmation activity: *Students follow provided instructions for an activity designed to teach a concept or confirm a principle.*			
Structured inquiry: *Students use designated procedures to investigate a question provided by the teacher.*			✓
Guided inquiry: *Students design their own procedures to address a question provided by the teacher.*		✓	✓
Open inquiry: *Students pose their own research question, design an experiment, and decide how to analyze their results.*	✓	✓	✓

Teacher directed ⬍ *Student directed*

Figure 12.1. Levels of inquiry. Check marks indicate student control over various steps in conducting an investigation.

ronment for learning science as science is practiced is challenging for both teachers and students (Crawford 2000; Trumbull et al. 2005; Trumbull et al. 2006). Teachers who have never had the opportunity to carry out their own research understandably feel ill prepared to facilitate investigations by their students. Citizen science, with its connections to professional scientists and authentic research, can help bridge this gap. Our focus here is on ways in which citizen science can intersect with classroom science, with particular attention to aspects that give teachers the knowledge, skills, curricular resources, and confidence needed to fully engage students in all aspects of research.

Curricular Support for Inquiry in Citizen Science

While some citizen science projects were created specifically for classroom use, many projects designed for public use also provide support for teachers who would like to implement them with their students (Table 12.1). One form of support is curricular resources that assist teachers and students in

TABLE 12.1.
Examples of citizen science projects that provide supports for classroom use

Project	Classroom focus	Inquiry components
	Designed for K–12 student use	
Project GLOBE (Global Learning and Observations to Benefit the Environment) globe.gov *International classrooms use protocols related to atmosphere, hydrology, soils, land cover, and phenology.*	Transitioned in 2005 from emphasis on scientists publishing research using student-collected data to scientists connecting their research to the interests and learning goals of students and teachers (Penuel et al. 2006).	"New Generation GLOBE," which supports teachers in helping their students understand science and do their own research projects (globe.gov/science).
Journey North www.learner.org/jnorth *U.S. and Canadian classrooms study wildlife migration and seasonal change.*	Lesson plans and activities related to Journey North's individual species and projects, such as Bald Eagles, leaf out, and photoperiod.	An online collection of resources that addresses inquiry strategies, such as how to pose questions that inspire scientific thinking and what types of questions help students critically review research by their peers (www.learner.org/jnorth/tm/inquiry/menu.html).
Forest Watch www.forestwatch.sr.unh.edu *Over 160 schools across New England monitor tree health and atmospheric ozone.*	Rigorous scientific protocols for biological and geophysical assessments, working both in the field and in the lab.	Opportunity for teachers to be mentored by scientists in research methods.
GREEN (Global Rivers Environmental Education Network) www.earthforce.org/GREEN *GREEN schools around the world each seek their own local partnership with a science institution.*	Collection and analysis of data to address a locally relevant question or problem.	Partnering with local scientists to help students conduct research; community action projects of students' own design that are integral to GREEN.

(Continued)

(TABLE 12.1—Cont.)

Project	Classroom focus	Inquiry components
	Designed for public use but include supplemental K–12 materials	
Project BudBurst www.neoninc.org/budburst *Participants record seasonal phenology of plants.*	Learning about plant observation and plant physiology, and connecting phenology observations with data on climate change.	An online collection of lesson plans, e.g., using BudBurst data to investigate how plant species might be affected by climate change (www.neoninc.org/budburst/educators/).
Monarch Larva Monitoring Project www.mlmp.org *Participants collect long-term data on larval monarch populations and milkweed habitat and build a better understanding of how and why monarch populations vary over time and geographic setting.*	Extensive K–12 curriculum resources provided by the complementary **Monarchs in the Classroom** project, including age-specific lessons that cover monarch life cycles, systematics, ecology, conservation, and migration (www.monarchlab.org/mitc).	Inquiry-based K–12 lesson plans on science content and processes, and workshops for teachers (www.monarchlab.org/mitc/resources/LessonPlans).
eBird www.ebird.org *Participants keep track of the birds they see anywhere in North America.*	Lessons and materials for teaching about bird diversity and identification provided by a complementary **BirdSleuth** curriculum module, *Most Wanted Birds*.	Free lesson plans in BirdSleuth's *Investigating Evidence* module guide each step of research: asking a scientific question, collecting data or evidence, drawing conclusions, peer reviewing, and publishing results (www.birds.cornell.edu/birdsleuth/inquiry-resources).

going beyond data collection to more fully experience the research process. Journey North, for example, provides online resources to guide teachers in facilitating all aspects of student investigations focusing on wildlife migration and seasonal change. Through this project, students collect data on selected animal sightings, plant budding, changes in daylight, or other phenological events. The Journey North lesson plans and teacher guides include an online collection of inquiry strategies, such as how to pose questions that inspire scientific thinking and what types of questions help students critically review research by their peers.

At the Cornell Lab of Ornithology, the BirdSleuth curriculum was designed to support student participation in citizen science projects, primarily at the middle-school level. The opening module covers how to conduct observations, identify birds, and enter data into eBird, a vast citizen science database maintained by the Cornell Lab. It also guides teachers in how to use the Lab's databases to discover and present trends and relationships in data collected regionally or nationally. At the heart of BirdSleuth is an online module called *Investigating Evidence* that focuses entirely on helping teachers guide students through independent inquiry. Whether using their own bird observations or observing patterns in online data collected by others, BirdSleuth students are encouraged to develop their own research questions, then design and conduct an investigation addressing a question of their choice. Depending on the type of question, such investigations can involve observational studies, experiments, or query and analysis of data output from appropriate citizen science databases.

The University of Minnesota's Monarchs in the Classroom project similarly offers K–12 lessons on how to conduct experiments along with age-specific lessons covering life cycles, systematics, ecology, conservation, and migration of monarch butterflies and other insects. Monarchs in the Classroom also supports student involvement in the Monarch Larva Monitoring Project, developed by researchers at the University of Minnesota to collect long-term data on larval monarch populations and milkweed habitat and to build a better understanding of how and why monarch populations vary over time and geographic setting.

Rationale for Inquiry

The focus on inquiry by citizen science project developers, beginning about 2005, represents evolution in the field. Earlier efforts to support classroom learning through citizen science focused primarily on student involvement in data collection, with the presumption that an understanding of scientific processes would naturally follow. Citizen science activities did provide authentic contexts for learning, motivating students to collect data carefully because of the importance of this effort to a larger scientific endeavor

(Means 1998; Rock and Lauten 1996; Tudor and Dvornich 2001; Fougere 1998; Evans et al. 2001). Partnerships between schools and scientists succeeded in demonstrating student proficiencies in data collection across various topics and taxa (Becker et al. 1998; Congalton and Becker 1997; Murphy 1998). And peer-reviewed articles incorporating student-collected data have been published (e.g., Hiemstra et al. 2006; Robin et al. 2005; Verbyla 2001), despite potential resistance to the idea that student-derived data could be sufficiently accurate for publication (Brewer 2002; Harnik and Ross 2003b).

The downside of rigorous data-collection efforts was that they tended to place disproportionate emphasis on data and thereby disconnected students from the overall research (Berkowitz 1996; Means 1998; Moss et al. 1998). Teachers' reluctance to prioritize routine data collection over more substantive learning opportunities resulted in limited citizen science participation by students and irregular data collection for researchers (Penuel and Means 2004). In addressing such challenges, projects have realized the need to move beyond data collection to also support student inquiry. Rather than treating students merely as data collectors, scientists can make their projects fulfill the needs of schools by helping teachers incorporate data collection efforts into student investigations. A study of GLOBE teachers, for example, demonstrated that classes were more likely to collect and contribute data regularly and reliably if their teachers were supported in developing student investigations (Penuel and Means 2004). Similarly, Shirk (2004) presented preliminary evidence indicating that integrating an inquiry component into protocol training can heighten high-school students' interest in the research question and increase their motivation to carefully collect project data.

Data collection can serve as a springboard to generate student questions that can lead to investigations. Summarizing research on implementation of a field-test version of the BirdSleuth curriculum with three teachers and their fifth- through eighth-grade students, Tomasek (2006) described the students' participation in eBird as a "question engine—an activity that engages students in making observations and inferences as a precursor to generating research questions" (206). Students reported that their research questions most often were based on their own curiosity as opposed to having been presented by the teacher or in curriculum materials. The three teachers in Tomasek's case study each related how students developed questions through watching birds.

Although inquiry offers a promising strategy for involving students in citizen science beyond the data-collection phase, trade-offs will continue to exist between goals for education and for scientific research. Project GLOBE, for example, has found school partnerships difficult to sustain unless they address the need for inquiry support, but GLOBE's move to support inquiry has shifted its focus from global-scale efforts to more targeted research.

Journey North offers robust materials for supporting classroom inquiry, but it is unclear whether project data are used for actual scientific studies. In Forest Watch, the balance between student inquiry and professional research depends on classroom-specific implementation and opportunities to connect with project scientists. In BirdSleuth, students contributing to eBird or another of the Cornell Lab's citizen science databases have no direct connection with the scientists but do have access to the same data outputs and visualizations used by professionals and other members of the public.

Nurturing Inquiry through Citizen Science

In addition to curricular resources, citizen science projects can provide a variety of materials and opportunities in support of inquiry-based learning.

Data Displays and Analysis Tools

Providing user-friendly ways to view and analyze data submitted by participants is a key way in which projects can support classroom inquiry. Teachers might choose to have their students view such data to introduce the project, perhaps generating discussion about potential research questions before the students collect any data of their own. After collecting data, students can view the larger data set to put their findings into a broader context over time or geographic setting or both. For example, students can query the eBird database to determine which birds are currently being seen in their area or to investigate questions such as which species migrate and which do not. They might choose to view trends in sightings of a specific species over a period of several years, or to determine the geographic range of a species that is seen in their town only rarely. Viewing outputs such as these provides opportunities for students to observe patterns and trends, develop inferences, and discuss various interpretations of the data.

Teacher Professional Development

Providing teacher professional development is another way in which citizen science projects can support student inquiry. Follow-up with twenty secondary teachers involved in the Monarch Larva Monitoring Project found that the number who provided opportunities for their students to conduct full inquiry increased significantly following two weeklong workshops in which they received intense instruction in each step of inquiry (Jeanpierre et al. 2005).

The Cornell Lab of Ornithology has been piloting a 5-week Web-based course designed to enhance the ability of middle-school teachers to engage their students in investigations focusing on birds. Using the same BirdSleuth readings and activity sheets designed for student use, teachers design and conduct investigations. After writing simple research reports they exchange peer reviews, revise their writing, and submit final reports summarizing their investigations. Online discussions throughout the course provide a forum in which teachers exchange advice and reflect on challenges and re-wards of implementing student research in their classrooms. For example, teachers select and apply a rubric to grade sample research reports, and they reflect on their approach to inquiry teaching as viewed on a continuum from structured to open inquiry (Figure 12.1). Pre- and post-surveys indicate movement toward open inquiry in teachers' intentions for the upcoming school year. Teachers also grew in their perceived level of expertise in helping students conduct investigations and analyze, interpret, and report their results. Teachers sometimes struggled with unexpected results in their investigations, and fellow course participants offered advice and alternative solutions. For example, one teacher commented, "This is usually when I stress to my students that part of science is sometimes not gathering as much data or the data that we would expect. At least now we know this before doing it with our students, right?"

Mentoring by Scientists

A teacher's success with innovative teaching through inquiry and citizen science can hinge on the level of external support received, including help from partners in the scientific community (Brewer 2002; Drayton and Falk 2006; Evans et al. 2001; Penuel et al. 2005). Types of support that scientists can provide on a local scale include mentoring on how to generate interesting research questions and how to analyze and interpret results. Research suggests that teachers who work in close relationships with scientists develop confidence in their ability to facilitate student inquiry (Trautmann and MaKinster 2005).

Teachers who are not comfortable with helping students pose research questions can be supported by scientists who provide solid content knowledge and a guided process for refining curiosities into questions (Trumbull et al. 2005). Scientist partners are well positioned to provide background information, standardized protocols, and support for implementing protocols in the field. Ideally, teachers introduce students to the research goals along with relevant protocols for data collection and analysis. Once students become familiar with the overall research, scientists or teachers or both can work with them to refine their own testable questions that utilize data collected with the standard protocols. With this approach, students collect data for scientists while at the same time addressing their own research questions.

On a broader scale, scientists can help teachers implement student investigations by providing curriculum materials and professional development opportunities. Viewing teachers as partners rather than recipients of outreach, scientists and project developers can target the trainings, materials, and support they provide to best meet teachers' expectations and needs (Penuel et al. 2007; Trumbull et al. 2005). Projects are most likely to enhance overall learning outcomes when they align inquiry-teaching activities with curriculum standards for content learning (Krajcik et al. 2008). Key features of professional development that supports inquiry appear to be provision of deep content information, training in how to employ the desired research protocols, time for open exploration in the area of research, and facilitation of an understanding of scientific questioning and the choosing of student questions (Jeanpierre et al. 2005).

Student Publications and Events

Another way in which citizen science projects can help students experience the full research experience is by providing a forum for presenting results. Monarchs in the Classroom, for example, holds an annual insect fair to spotlight student research. In BirdSleuth, students can submit written research reports to the Cornell Lab of Ornithology for publication in a semiannual publication called *Classroom BirdScope*.

Successful implementation of citizen science in school settings requires attention to the potentially competing needs of students, teachers, and professional scientists. Scientific rigor cannot come at the sacrifice of student learning, or vice versa. If, however, standardized protocols are used as a platform for student investigations rather than just data reporting, goals for both students and scientists can be achieved. Students benefit through use of rigorous and well-tested protocols that are provided and supported by scientists, and scientists benefit through access to student-generated data. Additional benefits for students include motivation, relevancy, and opportunities to develop scientific skills along with visions of what it means to be a scientist.

Scientists who invest in helping students conduct open-ended inquiry projects can potentially improve their own research outcomes in two ways: first, by direct improvement in terms of data quality, in cases where inquiry inspires more careful data collection; and second, by indirect improvement in terms of sustained partnerships, in cases where supporting inquiry meets key curriculum requirements for participating teachers. Additional benefits for scientists include opportunities to positively influence classroom learning, raise awareness about target species or topics, and cultivate the next generation of citizens.

Acknowledgments
Many of the projects discussed in this chapter have been supported by the National Science Foundation (NSF), but any opinions, findings, and conclusions are those of the authors and do not necessarily reflect the views of NSF. We wish to thank the many teachers around the country who have collaborated with us in field testing materials and seeking the best fit between inquiry and citizen science in classroom settings.

13

A Gateway to Science for All

Celebrate Urban Birds

KAREN PURCELL, CECILIA GARIBAY, AND JANIS L. DICKINSON

How do we achieve an equitable representation of all audiences in citizen science in general and birding in particular, and why is it important to do so? In this chapter we explore these questions and take a closer look at successful approaches and project design features to broaden participation. We use Celebrate Urban Birds, a citizen science project developed by the Cornell Lab of Ornithology in 2007, to highlight strategies to include audiences who may not have had opportunities to connect with nature or to see themselves as science participants.

Early surveys of citizen science projects at the Cornell Lab of Ornithology were disappointing with respect to diversity; participants in the majority of projects were highly educated, upper-middle class, middle-aged or older, and white. This is not surprising; according to the U.S. Fish and Wildlife Service report *Addendum to the 2006 National Survey of Fishing, Hunting, and Wildlife-Associated Recreation* (U.S. Department of the Interior 2009), birders are neither racially nor ethnically diverse.

The rarity of ethnic minorities in environmental science is notable, and amateur ornithology, or "bird-watching," certainly reflects this. Serious bird-watchers who remain active for 20 years or more will meet fewer than three African American bird-watchers (Robinson 2005). Exploring the virtual absence of African Americans from this popular hobby, John C. Robinson acknowledges that, in spite of the dearth of other African American birders and given his own apparent talent for identifying birds, he came to a point when he consciously "gave himself permission" to join the club, becoming an avowed bird-watcher and eventually a professional ornithologist with experience in both the public and private sectors.

Clearly, if we are going to achieve the broader goals of citizen science described in the introduction to this book, an effort must be made to bring equity to citizen science by engaging diverse audiences. In this chapter, we discuss the role that citizen science can play in addressing inclusion in nature study and in developing awareness and interest in biodiversity and environmental science.

Why Diversity Matters

Working to achieve a more equitable representation of all audiences in citizen science makes good sense on several levels. Inclusiveness not only promotes scientific literacy across cultures, thus engaging a larger pool of talents and viewpoints in science and technology, it also ensures that citizen science data represent the true variation in cultures, ethnicities, and geographical distributions. This has both scientific and ethical implications. Scientifically, if environmental justice communities are not sampled, the data will paint a rosier picture of environmental quality than actually exists. Ethically, the people who are most vulnerable to environmental injustice deserve to have access to information, community support, and educational empowerment to make their voices heard (Higgins 1993). It is difficult to imagine how this can come about without scientific empowerment and, in the case of urban audiences, early opportunities to discover and embrace nature and the environmental sciences. Uriarte and collaborators (2007), in fact, proposed that our current value system and culture of science promote a narrow set of outcomes, restrict the diversity and relevance of scientific pursuits, and lower the level of public engagement. They suggest that activities applying scientific knowledge to policy or to improving the everyday lives of the public are essential to promoting change.

But how do projects expand their audiences? Projects constructed under a majority worldview may have unseen biases that unwittingly exclude cultures that don't match the dominant culture's way of looking at the world. Subtle expectations in the language of science, approaches to learning, and discourse may create an environment that is incongruent and exclusionary (National Research Council 2009). Expanding audience diversity in citizen science is a challenging undertaking because some minority groups feel alienated and disconnected from science beginning as early as elementary school (Kahle and Meece 1994). Although family-based programs that include active parental involvement seem to improve the confidence and achievement of students from groups historically underserved in the sciences (Smith and Hausafus 1998), barriers (e.g., lack of transportation, time, money, or language proficiency) and other social factors (safety concerns, lack of self-confidence, lack of role models, fear) may prevent willing families from participating. Cultural factors may come into play as well. For example, some researchers suggest that African Americans may be more interested in participating in projects that offer opportunities for leadership and empowerment and tend to be more invested in environmental justice and change for their communities than in contributing to mainstream endeavors or more general conservation efforts (Arp and Kenny 1996; Johnson et al. 2004; Parker and McDonough 1999).

Some studies suggest that the primary causes of hesitancy or unwillingness to join scientific and environmental organizations are differing worldviews and lack of trust (Hassel 2004). Potential participants may not see themselves as individuals who *can* contribute to the scientific community or as people who *belong* in the citizen science community, or they may not fully trust this community as a safe place to try new things or make mistakes. Although lack of trust can easily be misconstrued as lack of interest, empirical data suggest that the interest level is actually quite high. Focus groups conducted with Latino communities in four major cities in the United States, for example, showed that adult participants viewed science positively and generally believed that it contributes to society (Garibay 2009). In addition, adults easily identified potential benefits of participating in citizen science, both for themselves and their children.

On the other hand, consistent with prior research indicating that youth attitudes toward science are determined by their in-school learning (Lawrenz 1976), many youth participants reported that science classes were boring and they expressed little interest in pursuing science. One youth said candidly, "[Scientists] are a little boring; they are just focused on their experiments." Although the focus group participants saw the importance of science, most could not see themselves as scientists and felt disconnected from science even to the point of failing to recognize that their own interests were scientific: "Science is okay, but we're just not that into it. Like, I really want to be a veterinarian." Adult participants in the focus group, while noting the value of science, still saw scientists as highly devoted to their work, isolated from friends and family, or "geeky" (Garibay 2009). Neither adults nor youth saw themselves as scientists or bird-watchers, and none felt a part of these communities of practice.

Celebrate Urban Birds

Celebrate Urban Birds (CUBs), developed and launched by the Cornell Lab of Ornithology in 2007, investigates new ways of creating centralized citizen science projects that reach diverse urban audiences. Here we explore the thought processes and specific strategies we used in creating CUBs to reach people who may have tenuous connections to science and nature and who do not already participate in science or scientific investigation. By describing these strategies and reporting on the results and achievements of the project, we hope to foster an approach that values inclusiveness and moves us closer to developing citizen science projects that are a good fit to a broader diversity of participants. Based on our experiences with CUBs, we believe that inclusiveness requires straightforward but flexible project designs, multiple entry points, strong community support networks, and a shared power structure. As CUBs demonstrates, projects that allow creative modification by practitioners can readily be brought into an assortment of organizations and across national boundaries.

Goals of CUBs

Citizen science balances a broad mix of scientific, educational, behavioral, institutional, and community goals. For a project to be successful each of these goals must be considered and weighed in the initial design. From its inception, CUBs was designed to form learning partnerships with underserved communities and to create an engaging scientific experience with a simple but rigorous protocol. Equally important in creating a successful project is considering goals of communities and community-based organizations. True collaborations with local organizations, neighborhood engagement, youth development, and learning are of primary concern and require ongoing bi-directional dialogue.

CUBs combines strong collaborations with community-based organizations using a simple project design. It attracts participants unfamiliar with birds and answers questions of interest within the newly emerging field of urban ecology. The scientific goal of the project is to determine how green space and the surrounding landscape influence distributions of birds in cities. Additional goals are to investigate how the quality, size, and arrangement of urban vegetation patches influence occupancy within urban areas and across the urban-to-rural gradient. Participants select a small bird-watching area, look for the presence or absence of sixteen species of birds for 10 minutes, and answer a few questions about habitat. We ask that they observe their bird-watching area 3 times to allow for estimates of detection probability (see Chapter 6).

Partnering organizations and their corresponding communities have their own goals, including to improve habitat, provide quality educational opportunities for families, address health concerns, promote community engagement, involve parents in their children's education, provide safe and meaningful leisure activities for families, and connect families with nature. Collaboratively addressing both the scientific and community goals is essential. This means that partners use their expertise to recruit and support participants, creatively drawing on the strengths of local neighborhoods to design "events" that bring people together to learn about birds, collect data, improve habitat, and create a sense of well-being, while the Cornell Lab provides engaging resources, educational materials, and expertise in the areas of science and nature, and creates a supportive and welcoming bilingual (Spanish and English) network of support. These collaborations create local communities of practice to support both scientific and community goals.

Our collective learning goals target a broad age range from kids to adults and elders. They include learning bird identification and observation skills; scientific concepts (e.g., bird-habitat-landscape relationships); new ways of viewing science (science as fun, interesting, connected to community, something "I" can do); and leadership (the ability to transmit interest and skills to a group). In 2009, the National Research Council (2009) highlighted two promising design features for projects to better support science learning among nondominant audiences. These include codeveloping projects with interests and concerns of the communities in mind and tailoring the educational design to strengthen community and family support networks. CUBs has been developed with an understanding that its growth depends on an iterative, ongoing collaboration and connection with community-based partners. Online interaction, phone conversations, events, webinars, workshops, and a willingness to listen and grow have been key to creating local and networked communities of practice from the project's inception.

Design Features

Simple and Flexible

One key to the success of Celebrate Urban Birds is a design that allows community partners to "drop it" easily into their existing agenda. The protocol is simple: participants are asked to use the free educational kit to learn about the sixteen focal species. The bilingual kit includes information about each of the species including silhouettes, drawings, and paintings as well as data forms, instructions, and seeds for planting sunflowers to attract birds. After learning about the birds, participants decide when and where they will watch them. They are encouraged to choose a variety of areas to observe, for instance, parking lots, green balconies, tree-lined streets, and parks. Participants select a defined bird-watching area of approximately 50 feet by 50 feet and answer simple questions about the habitat. They watch their selected area for 10 minutes and record the presence or absence of the sixteen focal species of birds. They are encouraged (but not required) to observe birds in their area at least 3 additional times during the same week. Participants have the opportunity to check the response labeled "unsure" when they are not confident of the identification of a particular species. Those who are not familiar with birds but wish to participate or lead others in participation can begin by focusing on only one or two species and add species as their skill level increases. People can participate as individuals or with others. Data may be submitted on scannable paper forms that are mailed to the Cornell Lab or entered into an online data-entry system. In addition, data submitted can be viewed online and are summarized with pie diagrams. Celebrate Urban Birds provides additional support to community-based organizations by offering free bilingual training webinars, video conferences, phone and e-mail support, and mini-grants. The project's flexibility and strong bilingual support allow group leaders with no experience in outdoor programming to easily participate and lead others in participating. Leaders have everything they need to make the project work for their community.

Shared Power Structure

Our research to better understand Latino community needs for citizen science indicates that community-based partners in placed-at-risk Latino communities face many complexities and concerns that may make participation in citizen science projects challenging. Interviews with group leaders in six major urban centers brought common issues to the surface: lack of staffing, not having "enough knowledge" to participate in citizen science, and shortage of resources, materials, and technical expertise (Garibay 2009). It is important to consider community realities such as language and literacy, work schedules, and relevancy to the local community. "Sincerity" was mentioned as a key component in successful partnerships with placed-at-risk Latino communities. For instance, one group leader working with inner-city youth said, "We partnered with a college that was not in our community. Our kids felt that the college students were participating just for the grades. There was no sincerity." Consistency and a sense of completion were also important factors.

Building relationships and sharing mutual goals are key to success: we have dynamic, ongoing conversations with community organizations, highlight their strengths, and collaboratively address concerns. The goals of Celebrate Urban Birds stem from a belief that learning is a cultural process deeply entrenched in history and placed within the local community (Nasir et al. 2006). We started by connecting with local organizations that reflect the strengths, concerns, interests, and motivations of local communities. These community-based organizations create a locally supportive atmosphere for the public to participate in citizen science. Organizations create festive "events" to learn about birds and citizen science and they use the arts, gardening, and community service to engage the public. Organizations recruit participants and embed the citizen science activity in a context that makes sense locally. They may create a block party focused on improving local habitat with youth leaders teaching families about birds and citizen science, or they may integrate the project into a church program for families recovering from abuse. A soup kitchen may focus on creating a community garden monitored by homeless guests doing a weekly survey of the sixteen species, or a local YMCA may create a program focused on the arts, birds, and citizen science at a waiting room in a local courthouse, creating a safe place where families can do something productive and soothing.

The range of partners has grown to include faith-based groups, libraries, community centers, wellness programs, youth groups, schools, after-school programs, battered women's shelters, senior centers, and rehab programs (see Chapter 4). These organizations interact with different age groups in diverse contexts (health, education, entertainment, the arts), and 80% work with underserved communities. Working with partners that are trusted, familiar, and embedded within local communities helps build institutional trust (Garibay 2004). Participants reluctant to engage with the Cornell Lab of Ornithology may feel comfortable and secure within the structure and familiarity of the local community group leader and organization.

In asking the community to create events and other activities, our design emphasizes the strengths and values of the community rather than its deficits (Bouillion and Gomez 2001; González and Moll 2002; Yosso 2005). Based on feedback from the community, we actively promote activities, challenges, or outdoor "quests" and provide support materials, including ideas developed by partners, generating an ever-expanding set of Web resources that provide multiple points of entry into the project. These resources include feature pages focused on how our communities have used the arts, gardening, science, mentoring, and urban greening to get people outdoors and teach bird identification, bird biology, habitat improvement, and more (Figure 13.1). The key here is that the project draws on participants' dynamic and personal experiences within the context of their own families and communities. For example, those who are comfortable in the arts can use the arts to teach others about bird identification; people who are exciting storytellers create engaging stories about bird behavior; dancers can "show" specific bird movements; gardeners may teach others how to grow and cultivate plants that are good for birds and show creative ways to improve habitat; and musicians may bring birdcalls and birdsong to the forefront. These approaches lead to "knowledge that is useful, powerful, and transferable," empowering participants to take action and become active contributors to their emerging "science community" (Basu and Calabrese Barton 2007).

The Celebrate Urban Birds mini-grant program has also contributed significantly to the project's success in engaging community partners. In applying for a grant, communities plan out their events (blending habitat improvement, the arts, birds, citizen science, and community involvement) and find arts and gardening partners. We collaborate not only with the official mini-grant winners, but also with the other 600-plus organizations that apply, providing free educational materials and Celebrate Urban Birds kits as well as tools, such as bilingual training webinars, model press releases, promotional flyers that can be personalized, video conferences, pho-

Figure 13.1. Students display their 3-D bird figures at Pine Hills Elementary School in Albany, NY. Thanks to a Celebrate Urban Birds mini-grant, staff, students, and their families created bird and butterfly gardens, collected data, participated in music and art activities, and put up nest boxes.

tos, PowerPoint presentations, and certificates of appreciation to facilitate their work. This feature of the program results in many more organizations fulfilling their visions than the twenty that receive grants annually.

Multiple Points of Entry and Strong Community Networks
A key component of creating a welcoming atmosphere in citizen science projects may be the creation of collaborative communities of practice among people who do not necessarily see themselves as participating in science activities or as leading others in science activities, but who respond to an invitation to "do" science through observation or experiences connected to their everyday lives. Rahm (2002) explored the effectiveness of an inner-city gardening program in the creation of an "emerging community of practice" that integrates work, science, and community values. Such an approach highlights the value of an experience that emerges from "doing" science that is meaningful, real, and relevant (Calabrese Barton 2007), in this instance, in a garden setting.

In this way, Celebrate Urban Birds is not just about collecting data. We encourage multiple experiences "doing" science, whether through meaningful observation, community improvement, artistic means (dance, visual arts), gardening, health, education, conservation, community, or insightful inquiry. These multiple entry points merge citizen science with participants' existing experiences and comfort zones. Two particularly strong gateways that place science in the context of the community are the arts and gardening. Jacobson (2007) says that "promoting conservation through the arts may reach a more diverse audience and reach them more successfully by engaging their hearts as well as minds." The arts and gardening also provide a colorful, showy way to attract attention to the project, promoting participation, self-expression, ownership, sense of place (Sobel 2004), leadership, and authentic, meaningful experiences.

Engaging families as the target audience is also a powerful strategy for working with urban communities. By providing multigenerational materials, removing language and literacy barriers, and encouraging multiple points of entry into the project, we create conditions that allow any combination of family members to participate and talk about and share their experiences. Calabrese Barton and colleagues (2001) pointed out that mothers who "do" science in familiar contexts with their children are more likely to have a dynamic and inquiry-based view of science. "Mothers seemed eager to help their children understand how the natural world around them works and were quite competent at transmitting both content and process of everyday science to their children." This research was instrumental in guiding our focus on families embedded within communities.

Curiosity and dialogue are the natural beginnings of scientific inquiry, and as in professional science, learning takes place whether or not things work out as planned. Sometimes sharing negative experiences or even fears about birds leads to deeper engagement and understanding. We encourage and support peer networks by featuring community celebrations, issuing challenges, sharing participant stories on the website, and managing social networking pages on Facebook and Twitter. Seasonal challenges are quests to find or experience nature, leading participants to share their findings and take ownership of their scientific experiences. One example of a seasonal challenge is "Funky Nests in Funky Places," which expanded the participants' experiences beyond bird counts by asking them to seek out interesting nests and share artwork, photographs, or stories about their observations, giving authority to participants' voices and initiating genuine scientific dialogue.

Impacts

In the 4 years since its inception, Celebrate Urban Birds has partnered with over 6000 community-based organizations, distributed over 180,000 educational kits, awarded dozens of mini-grants, and held 9 successful outdoor challenges or quests. Over 80% of participating organizations serve underserved audiences. Our participants range from preschoolers and kindergartners to seniors, and 60% have little or no experience with birds. Informal Web surveys conducted by program staff indicate that individuals and organizations experience a range of learning outcomes, including learning how to identify new species, observing interesting bird behaviors, learning how to attract birds, and collecting data. In addition, organizations are taking part in varied and meaningful activities such as planting bird gardens, sponsoring art activities, setting up feeders, and encouraging others to watch birds, as well as teaching and learning about native plants and threats to birds in cities. Although it is a new project, evidence is growing rapidly about its effectiveness as a national citizen science effort that is inclusive and inviting to a broad diversity of participants.

National citizen science projects that are designed or, even better, codesigned with local communities' interests and values in mind are more likely to appeal to people from a broad range of cultural, ethnic, geographic, and economic backgrounds. We believe that this broad appeal is the core strength of CUBs. Strategies that have worked successfully for CUBs begin with setting clear objectives for reaching scientific, collaborative, and public learning goals and collaboratively addressing community and partner goals. Building strong partnerships with community-based organizations has also been vital. Design features that have contributed to the success of the project include creating a sense of fun; tapping into strengths, values, and knowledge of the community through a shared power structure; and maintaining strong community support networks by providing multiple gateways into science and the natural world. In addition, a simple and flexible approach has been important, fitting easily within the goals of community-based organizations and allowing families to participate within constrained schedules and literacy and economic boundaries.

Acknowledgments
Celebrate Urban Birds was developed at the Cornell Lab of Ornithology in collaboration with Cornell Cooperative Extension NYC. The project was created and augmented on grants to coauthor Janis L. Dickinson from Cornell Cooperative Extension (Smith-Lever) and the NSF (DRL-000187689). The U.S. Trust (anonymous donor), the Adelson Family Fund for Citizen Science, the Arthur A. Allen Citizen Science Endowment, and The Wallace Foundation provided additional support for staffing, kits, Spanish translation, and multimedia.

14

Children and Nature

Following the Trail to Environmental Attitudes and Behavior

NANCY M. WELLS AND KRISTI S. LEKIES

Since the early part of the twenty-first century, attention has focused on children's dwindling connection to the natural world (Louv 2005). Parents, educators, and policymakers are concerned that children may be spending less time outdoors and that disconnection from nature may have detrimental effects on children's health and functioning. Evidence suggests that worry for children who have limited nature access may be warranted. Residential window views of trees and vegetation as well as time spent outdoors in natural settings have been linked to concentration and other aspects of cognitive functioning (Wells 2000) as well as to reduced symptoms of attention-deficit disorder (Kuo and Faber Taylor 2004; Faber Taylor and Kuo 2009), greater psychological resilience in the face of adversity and life stressors (Wells and Evans 2003), and lower rates of weight gain due to greater levels of physical activity associated with time spent outdoors versus indoors (Bell et al. 2008).

While mounting evidence links children's nature access to their health and functioning, impacts on development of children's interest in and commitment to the welfare of the natural environment are less well understood. Does a child's time in natural settings truly affect environmental attitudes and behaviors? What does research suggest about the short-term influence of programs or activities intended to connect children with the natural environment? Does a child's participation in a weeklong nature immersion or environmental education program, for example, yield immediate increases in environmental attitudes and behaviors? If so, are particular types of outdoor programs more effective? And in the longer term, is a child who grows up in an urban setting with a view of a lush, green park, likely to advocate for the creation and protection of urban parks later in life? To what extent might environmental attitudes instilled by childhood experiences be sustained over a lifetime? Does the amount or quality of time (or both) spent in nature during childhood ultimately affect individuals' long-term environmental attitudes and behaviors?

These questions are relevant to citizen science for a variety of reasons. Involving children in activities such as recording cricket calls, counting bird eggs, or monitoring monarch larvae may yield multiple benefits (Bonney, Cooper, et al. 2009; Cohn 2008). In addition to assisting scientists in the collection of data to document the status or movement of species and bolstering children's health-related outcomes by increasing time spent outdoors, such programs may enhance children's environmental attitudes and behaviors, lead to participants' lifelong dedication to environmental stewardship, and thereby ultimately contribute to global welfare over the long term.

This chapter provides an overview and a critical evaluation of literature examining linkages between children's nature experiences and their environmental attitudes and behaviors. We summarize the findings and critically assess the strength of the evidence provided by the extant research. Last, we suggest directions and strategies for future research and consider implications for citizen science.

A Brief Review of the Literature

This literature review is organized into two major sections. The first examines evidence concerning the short-term impacts of children's nature experiences. These are studies conducted in a short time frame, looking at the influence of children's exposure to nature on environmental attitudes and behaviors. In the second section of this chapter, we examine longer-term effects of nature exposure to consider hypothesized linkages between childhood nature experiences and later life environmental commitment or behavior.

Do Nature Activities Affect Children's Short-Term Environmental Attitudes and Behaviors?

Most studies examining the effects of nature activities or environmental education programs on children's environmental attitudes and behaviors have focused on short-term effects. These frequently are measured immediately on completion of participation in an activity or program or shortly thereafter, but also may be measured as long as 6 months or 1 year later. The literature primarily consists of small-scale studies that address outcomes of individual programs, rather than looking across groups of similar programs.

We examined studies of time spent outdoors through school-based environmental education programs or activities, class trips, residential programs, summer camps and after-school programs. Programs ranged in duration from 1 day, in the case of school field trips, to an entire school year for after-school programs. The age of participants ranged from 6 through 18 years, although most participants were children in the 9–12 age group. The participants engaged in a variety of environmentally focused activities including lab work, field observations of plants and animals, watershed monitoring, and soil experiments; they planted trees and flowers, recycled, gardened, learned about forest ecology, and went hiking, camping, canoeing, and skiing.

Comparison Studies

Some studies compared participants with a similar group of nonparticipants, or those in one program with those in another. These studies were almost exclusively quantitative using survey methodology, although some studies incorporated qualitative components as well, such as drawings, observations, and interviews with children, parents, or teachers (e.g., Palmberg and Kuru 2000). Overall, the impacts of participation in environmental education programs are positive, with the vast majority of studies indicating a greater improvement in environmental attitudes, behavior, or behavioral intention by children who have had some exposure to environmental education programs compared with children who have not (Bodzin 2008; Bogner 1998; Cronin-Jones 2000; Leeming et al. 1997; Skelly and Zajiek 1998).

The amount of direct contact with the natural environment may also make a difference. Research suggests stronger effects of longer residential environmental education activities compared with shorter ones (Bogner 1998; Stern et al. 2008), as well as for participants of residential programs compared with more limited classroom opportunities (Dettman-Easler and Pease 1999). Carrier (2009) found greater improvement in environmental attitudes among boys who had participated in outdoor schoolyard compared with indoor classroom environmental education activities, but no differences among girls in the two groups. Students in school-based outdoor educational and recreational activities expressed stronger concern for nature than students from schools with fewer such activities (Palmberg and Kuru 2000).

Noncomparison Studies

Noncomparison studies attempted to document the effects of environmental education or other nature-related programs without a comparison group of children. These studies either compared the same group of children over time—that is, before and after an educational program—or looked at effects on one group of children at just one point in time. Like the comparison studies, the results of these studies suggest positive changes in environmental attitudes and behaviors after participation in environmental education, residential, and summer camp programs. Research methods included surveys with participating children, parents, and teachers; observations; child and parent interviews; and use of children's drawings and journals. Evidence suggests that children who participate in such programs have greater interests in environmental issues, are more aware of and concerned about nature, want to learn more about nature, and are more likely to adopt pro-environmental behaviors such as recycling (Ballantyne et al. 2000; Dresner and Gill 1994; Farmer et al. 2007; Smith-Sebasto and Obenchain 2009).

What is much less discussed in the research literature is the reason these changes take place. Rickinson (2001:268), in an extensive review of environmental education research, indicated that "it is clear that there is some evidence to suggest that environmental education initiatives can effect changes in learners' environmental knowledge or attitudes. What is less clear, however, is how and why such initiatives bring about such effects." Attention to process has not been addressed in the research to the same extent as outcomes.

From studies that have addressed the processes or factors through which environmental attitudes are influenced, several key components emerge. One of the primary factors is the importance of active, hands-on activities (Ballantyne et al. 2000; Ballantyne et al. 2001; Bodzin 2008; Leeming et al. 1997). Other important factors include the addressing of local issues; involvement in individual or group projects, particularly those that involve information gathering or field research (Ballantyne et al. 2000; Bodzin 2008); the use of familiar and easily accessible sites, repeated exposure, discussions, and support for the teacher (Bodzin 2008); active engagement of teachers (Stern et al. 2008); sensory experiences that make interaction with nature more real and memorable; relationships with peers, instructors, and chaperones; novelty of objects, people, or experiences; and freedom to choose activities (James and Bixler 2008).

Citizen science programs, with their engagement of the public in meaningful scientific projects, demonstrate many of these key characteristics. Young people participate in hands-on activities, collect data in familiar field-based settings over multiple time periods, learn and discover new aspects of the natural world, work with others, and have opportunities to make genuine contributions (Bonney, Cooper, et al. 2009; Cohn 2008; Trumbull et al. 2000).

Strengths and weaknesses can be noted in the studies of short-term effects of nature programs and activities. A strength of this work is that among the quantitative studies, a control or comparison group was often used, which allows for more confident cause-and-effect conclusions (i.e., enhances "internal validity") (McDavid and Hawthorn 2006). In other words, there is more evidence that the changes in attitudes and behaviors were due to the specific environmental education activity rather than to some other cause. In addition, some studies used substantial sample sizes (McDavid and Hawthorn 2006), which also enhances our confidence that the nature program truly had an influence. And yet there are various other limitations in research design that might be improved.

A major methodological limitation in the cross-sectional and short-term pre-post research concerning children's exposure to nature relates to measurement. A variety of measures have been used, and the body of knowledge concerning the effects of nature on children's attitudes and behaviors would benefit from greater consistency of measures across studies. Greater consistency would allow for more comparability and would facilitate more coordinated national or regional analysis of programs. In addition, using established measures that have been tested in other settings would help bolster internal validity, enhancing confidence in the study's findings, as would more objective measures and reduced reliance on subjective measures. Also, most studies have focused on environmental attitudes; more effort should focus on environmental behaviors such as recycling, water conservation, and litter cleanup in both the long and short term. One example of a measure with established reliability and validity is the series of gamelike measures of children's environmental attitudes and behaviors developed by Evans and colleagues (2007). Using a felt board, puppets, and a game board, children are engaged in reporting their beliefs and behaviors. Environmental attitudes and values are measured with three games; a fourth measures environmental behaviors.

An additional shortcoming of most research examining the short-term influences of nature programs is that the studies typically focused on only one program, rather than looking comprehensively at several programs. Without replication, it is unclear whether similar results would be found for other groups of children at different points in time or with different adult leaders. Furthermore, posttests are typically given shortly after the end of the program with little examination of whether, and for how long, the changes remain. While some evidence suggests that that gains last 6 months or a year (Bogner 1998; Farmer et al. 2007), it is unclear whether the positive impacts fade with time (Stern et al. 2008).

Linking Childhood Nature Experiences with Later Life Outcomes

Various studies have aimed to link the experiences of childhood with environmental attitudes, behaviors, or activities of adulthood. In this section, we provide a brief overview of this research.

Significant Life Experiences Research

Among the earliest efforts to document an association between early life nature experiences and later life outcomes was the "significant life experiences" (SLEs) work. SLE research is not technically longitudinal in nature because data are not collected at multiple points in time. Rather, SLE uses autobiographical reminiscence data collected at one point in time to trace the origins of adult environmental commitment back to childhood influences.

In 1980, motivated to improve the efficacy of environmental education, Thomas Tanner published the first SLE study. Using a qualitative approach, Tanner (1980) asked staff of several environmental organizations (e.g., Sierra Club, Nature Conservancy, National Audubon Society, and National Wildlife Federation) to describe the early influences that led them to choose environmental conservation work. The responses were analyzed for content to reveal a variety of influences, including contact with various habitats, time in nature alone or with others, books, influential individuals, and the commercial development of a beloved open space. The most frequently mentioned influence, however, was time spent outdoors, cited by nearly all respondents. Moreover, among those who specified the influence of outdoor activities, hunting, fishing, and bird-watching were most often mentioned, although in some cases, hunting was associated with feelings of remorse and a decision to limit future participation in the activity. The majority of the influences occurred during childhood or adolescence.

Numerous subsequent studies have provided support for Tanner's findings (Birnbaum 2007; Chawla 1999; Corcoran 1999; Palmer 1993; Palmer and Suggate 1996; Palmer et al. 1999; Peterson and Hungerford 1981; Sward 1999). These studies, conducted across various countries including Australia, El Salvador, Norway, Canada, and the United States, have strikingly consistent findings. The greatest commonality among all findings is the importance of time spent outdoors in natural habitats during youth.

Whereas most of the SLE studies have focused on adults' reports of influential experiences, Arnold and colleagues (2009) employed the SLE approach to examine how and why youth environmental leaders (ages 16–19) became involved in environmental action. Consistent with the earlier studies of adults, the influential experience mentioned by all twelve youth participants was "time in nature."

The significant life experiences literature clearly suggests that childhood nature experiences are associated with later life environmentalism. This area of research is, however, quite controversial (Chawla 1998a, 1998b, 2001; A. Gough 1999; N. Gough 1999; S. Gough 1999) and has been subject to considerable criticism including allegations that studying environmental activists may not be desirable because activism may not actually yield environmental results; concerns that SLE findings may not clearly translate to environmental education curriculum development; complaints that the SLE findings are not replicable; claims that life experiences may not affect all people in the same way but instead are moderated by characteristics such as gender, age, and socioeconomic status; and questions regarding what SLE truly measures. Here, rather than attempting to adjudicate this debate, we focus briefly on issues related to internal validity, or the strength of the causal evidence, and external validity, or the extent to which findings can be generalized to other settings, contexts, groups, or time periods. We then discuss the strengths of SLE research.

First, with respect to external validity, because SLE studies typically focus on individuals who participate in environmental careers or activism, the findings may not be generalizable to the population as a whole, to individuals in other types of careers, or to people growing up in a different era. This concern arises whenever one specific, relatively narrow group is the subject of a study. For example, if a study is based only on young children, the findings are unlikely to apply to 30-year-olds; and findings from studies of low-income individuals may not generalize to high-income people. Similarly, because people in environmental careers may differ in significant ways from the general population or from people in other careers, we cannot assume the findings from research with this one group necessarily apply to people across careers or age cohorts.

Second, with respect to internal validity (i.e., confidence in cause-effect relations), the use of retrospective self-report or self-assessment weakens the measurement validity and consequently the internal validity of the study. In other words, measuring childhood time in nature based on individuals' own assessments is a subjective, potentially inaccurate strategy that is further jeopardized by the limitations and inaccuracies of memory. This makes causal direction ambiguous. Is it the case that certain childhood experiences lead people to be dedicated to the environment? Or, rather, do individuals construct coherent stories of their life experiences leading up to their ultimate environmental career or activism?

It is appropriate to note that advocates of the SLE approach typically embrace qualitative research based on a "phenomenological" framework in which subjective experience is valued and objectivity is not necessarily the goal. Thus they argue that it is not the "verity of memory" that is critical, but rather the "utility of memory." Chawla (1998b:388) states: "Researchers concerned about the verity of memory are preoccupied by questions such as, 'To what degree do errors enter the reconstruction of the past?' or, 'What conditions increase or decrease accuracy?' Those concerned with the utility of memory . . . ask a more phenomenologically rich set of questions. 'What is the meaning and use of memories during different periods of life?' 'How does the interpretation and use of memories reflect gender, class, culture, or other life conditions?'"

Thus, the applicability of the SLE approach may depend on what question is being asked. For studies aimed at establishing causal connections,

SLE may not be appropriate. SLE research may be most effective as a strategy to generate hypotheses or as a tool to be used in tandem with other methodologies.

Long-Term or Longitudinal Studies

In addition to the SLE studies, a few quantitative studies have examined long-term associations between childhood nature exposure and later life outcomes. These studies build on the foundation provided by the SLE work but differ from that work in several specific ways. First, these studies are typically quantitative—they involve the collection of numerical data rather than verbal interviews. Second, the variables may be subjectively measured through self-report or objectively measured using observational measures of behavior or the report of a teacher or relative. And, particularly notable, causal connections are established using statistical analyses rather than interviewees' experiences. These studies attempt to address some of the issues related to external validity (e.g., generalizability) and internal validity (e.g., cause-effect conclusions).

Lohr and Pearson-Mims (2005) surveyed over 2000 adults ages 18–90 living in major metropolitan areas in the United States to examine the association between childhood nature contact and adult attitudes regarding plants and trees. Childhood experiences such as planting trees and taking care of plants were among the significant predictors of favorable adulthood attitudes toward trees. Wells and Lekies (2006) examined linkages between youth nature experiences and subsequent adult environmental attitudes and behaviors, using data collected by Lohr and Pearson-Mims (2005). Findings suggest that engagement with both "domesticated" and "wild" nature in youth is associated with later life environmental attitudes (attitudes, for example, about the importance of trees to quality of life, the value of untouched natural areas, and the responsibility of humans to protect the environment) and behaviors such as recycling, volunteering to clean up natural areas, and choosing to spend time outdoors rather than indoors (Wells and Lekies 2006). A study of German adults found a modest association between childhood time spent in nature and adult "indignation about insufficient nature protection," which, in turn, predicted willingness to engage in behaviors to protect nature (Kals et al. 1999). Bixler et al. (2002) examined the association between childhood play and adolescents' environmental preferences. Adolescents who, as children, had played in wilderness areas, were more likely to prefer woodland paths than those who had played in the yard during childhood. In a related study, Ward Thompson and colleagues (2008) explored the association of childhood experiences in natural areas with adult use of and attitudes toward such places. Frequent childhood visits were associated with being prepared, in adulthood, to visit woodlands or green space alone as an adult. Individuals who did not visit such places in their youth reported a low likelihood of such visits in adulthood. Chipeniuk (1995) documented that people who reported having foraged the greatest breadth of objects (e.g., arrowheads, turtles, and berries) during childhood had greater knowledge of biodiversity as adolescents. In another study examining linkages between childhood and young adulthood, Ewert and colleagues (2005) found that the types of nature activities (e.g., appreciative, consumptive, media based) reportedly experienced during childhood predicted ecocentric versus anthropocentric beliefs during college.

By studying wider, more general samples of the population, these studies lend external validity to more typical SLE studies, which focus on environmental activists, professionals, and educators. In addition, the quantitative measures typically used arguably have greater measurement validity. These studies do, however, exhibit some of the same limitations evident in the SLE work. In particular, because the "long-term studies" often rely on retrospective self-report of childhood nature experiences, the validity of the measures is somewhat weak. Furthermore, because the studies are often not truly longitudinal—following individuals from youth to adulthood, with multiple points of data collection—temporal precedence is unclear, making causal conclusions tentative.

While the research briefly reviewed provides a variety of insights concerning the linkages between nature exposure and a range of outcomes, clearly there are many opportunities to improve on what has been done by conducting more rigorously designed research studies.

Future Research

Future research efforts should move toward more rigorous research design that will provide greater clarity regarding the causal linkages between environmental education or citizen science programs and participants' environmental attitudes, behaviors, and other outcomes of interest. Although studies may sometimes be carried out in the context of an organization whose primary mission is the delivery of programs, practitioner-researcher partnerships may yield straightforward strategies to increase the methodological rigor of the study without compromising the integrity of the citizen science or environmental education program. For example, to assess the efficacy of a program on participant outcomes, measures should be taken both before and after participation in the program. When possible, a comparison or control group should be used that does not take part in the program. Sometimes this can be achieved by creating a waiting list of individuals who eventually will participate. Ideally, to ensure that the groups are similar, individuals should be randomly assigned to either the wait list or the participant group.

Mediators

One fertile area for future research concerns mediators (Figure 14.1). Mediating mechanisms explain *how* or *why* one variable affects another. In this case, how or why might nature experiences such as citizen science or environmental education programs influence environmental attitudes and behaviors? Mediators are both theoretically and practically useful. Having greater understanding of cognitive or emotional *process*, or both, or of mechanisms of change (Rickinson 2001; Sibthorp et al. 2007) underlying attitude change or behavior change can provide both theoretical and practical insight to enhance the efficacy of citizen science, environmental education, and other intervention strategies. Mediating mechanisms provide leverage points that can be targeted, potentially through various programmatic strategies. For example, Chipeniuk (1995) provides insight regarding the possible mechanisms that might underlie an association between childhood foraging and later knowledge of environmental biodiversity. He suggests that the process of foraging facilitates cognitive adaptations that are ultimately manifested in greater environmental competence.

Moderators

Future research might provide more insight concerning what kinds of programs or interventions are most effective for whom and at what times. It is unlikely that one size will fit all potential participants in environmental education or citizen science programs; however, we know relatively little about what types of programs are most effective for what populations at what time of life. Moderators (or "interaction variables") tell us that the effect of a program or intervention on some outcome variable "depends on" some other variable (the moderator) such as age, gender (Carrier 2009; Zelezny et al. 2000), stage in life, or context (Figure 14.2). Greater clarity about which strategies are most effective for which age groups, which gender, and under what circumstances would allow for more nuanced and targeted citizen science and environmental education initiatives.

Figure 14.1. Mediators explain how or why a program has an effect on outcomes.

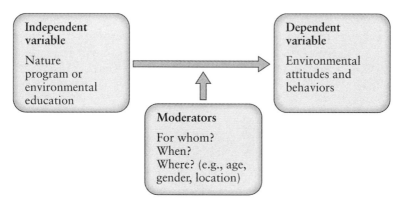

Figure 14.2. By including interaction effects, moderators address "it depends" questions of context and sample characteristics.

Common Measures across Studies

Last, greater comparability would be possible across studies of children's nature experiences if common measures were used by researchers in a variety of contexts. Some possible measures include the children's environmental attitudes and knowledge scale (CHEAKS) (Leeming et al. 1995), Bixler and Floyd's (1997) measure of fear of nature, and Evans's aforementioned interactive games to measure children's environmental attitudes and behaviors (Evans et al. 2007).

Implications for Citizen Science

The study of linkages between childhood nature experiences and environmental attitudes and behaviors is relevant to citizen science in multiple ways. First, it suggests that children represent a fundamental audience for citizen science programs seeking to have significant long-term impact on attitudes and behavior. Including citizen science as an independent variable, youth-oriented citizen science programs may be one potent strategy to engage children with the natural environment and to foster a connection with, knowledge of, and ultimately, we hope, a commitment to the natural environment. On the other hand, studies exploring the character of, the timing of, and the long- and short-term efficacy of children's exposure to nature can inform the design and implementation of citizen science programs targeting youth. For example, hypothetically, if research indicated that weeklong nature experiences involving intensive hands-on activities, and occurring between ages 8 and 12, were most likely to yield long-term effects on environmental attitudes and behaviors, then citizen science programs could be developed to meet such parameters.

One set of related issues concerns how to most effectively recruit children to citizen science programs and how to best ensure retention of participants in such programs. One possibility for recruitment is to partner with youth organizations for specific projects, or to incorporate citizen science activities into existing youth organization activities. Groups such as Scouts, 4-H, and after-school programs that include outdoor activities may be captive audiences for citizen science projects. Young people in such programs value new opportunities for learning, challenge, and enjoyment that provide them with new roles and responsibilities in which they can contribute and be taken seriously. These opportunities also foster an interest in long-term participation in youth programs (Ferrari et al. 2009). Citizen science, with its active, hands-on approach, can provide a wide array of novel activities to engage young people.

The use of technology in recruitment and retention of program participants is an area of some controversy. If children are increasingly drawn to, and spend time engaged with, computers and other technology, this may be a potent hook to connect youth with nature. Devices ranging from websites to cameras to Global Positioning Systems (GPSs) may bolster youths' enthusiasm for spending time outdoors. On the other hand, some may argue that technology is an adversary and that youth should be introduced to a "pure," technology-free nature experience. In fact, some research suggests an inverse relationship between environmental concern (e.g., regarding pollution) and trust in the efficacy of science and technology (Borden 1984/1985). This association reflects a belief that personal corrective action is not critical because science and technology will address the environmental challenge. Such linkages might be further examined in a contemporary context.

Another issue related to the recruitment and retention of youth participants concerns the activities in which young citizen scientists might become involved. How might the typical citizen science activities of observation, measurement, and reporting be most effectively presented and structured (or unstructured) to engage children and adolescents and maintain their involvement over some period? Supportive adult leaders can provide guidance to children and youth, create positive learning environments, and help young people become more comfortable with the natural environment (Ferrari et al. 2009). This is especially critical for youth who have little previous outdoor experience and may have strong fears and negative attitudes toward nature (Bixler et al. 1994; Bixler and Floyd 1997; see Chapter 13).

In addition to the potential role of citizen science programs as opportunities to engage youth in the natural environment, adult-focused citizen science programs may be related to this area of research as dependent variables. In other words, if youth engagement with nature is truly linked to adult environmental commitment, the future success of adult citizen science programs depends on current youth being exposed to and engaged with the natural environment. If levels of nature engagement among children dwindle, perhaps adult citizen science programs will falter in future decades. It is even possible that the overrepresentation of people over 65 in current citizen science projects reflects not only the free time of retired people but also a diminished connection to nature among middle-aged and younger audiences.

A particularly critical aim for future study is to gain a clearer understanding of what specific characteristics of children's participation in citizen science or other outdoor experiences are most likely to yield desirable outcomes. Researchers have examined various themes such as direct versus indirect contact with nature (Kellert 2002); active versus passive nature experiences (Lohr and Pearson-Mims 2005); and interaction with domesticated versus wild nature (Wells and Lekies 2006). An integrated taxonomy of children's nature interactions would add coherence and clarity to research efforts. Researchers might then more effectively examine the linkages between specific types of citizen science nature experiences and outcomes such as environmental attitudes and behaviors.

Citizen science programs can play a role in addressing a particularly interesting question concerning the "critical period" *when* nature exposure occurs. The hypothesis underlying studies of association between children's nature experience and adult attitudes or behaviors is that nature exposure during childhood is crucial to subsequent adult attitudes or behaviors. If it is true that children who miss their "earth period" or "bug period" fail to bond with nature (Hansen 1998), it may be possible to identify a specific "critical" period during which citizen science and other nature exposure is most influential.

Citizen science efforts can be integrated with other research efforts. This would allow for more types of data to be collected in tandem—data concerning natural species as well as data regarding the effects of children's outdoor experiences on their attitudes and behaviors—simultaneously enhancing our knowledge of natural species and broadening our understanding of the effects of outdoor experiences among youth. The literature regarding nature exposure on environmental attitudes and behaviors is dominated by environmental education research and is typically conducted within the context of school-based programs. Collaborations between citizen science investigators and other researchers would broaden the scope of research, enhance the rigor of research findings, enable the use of comparable measures across studies, and contribute to a more sophisticated and nuanced understanding of what works, how it works, and for whom it is effective.

Acknowledgments
Thanks to Kimberly Rollings and Sarah Lenkay for assistance with this chapter.

15

Internet-Based Social Networking and Collective Action Models of Citizen Science

Theory Meets Possibility

HEATHER A. TRIEZENBERG, BARBARA A. KNUTH,
Y. CONNIE YUAN, AND JANIS L. DICKINSON

The Internet has changed how people assemble information, how they interact, and the distance over which they can effectively collaborate. Activities that once were local have become both local and global with the ability to view local benefits within the broader contexts of interconnectedness, cumulative impacts, and scale. The growth of citizen science is an excellent example of this evolution with widespread use of the Internet to collect observations locally, submit them to a centralized database, interact with and display the collective data at larger geographic scales, and interact with others over observations and results. In addition to creating online learning communities and engaging people with the natural world, citizen science projects engage people in self-selected cooperative endeavors to address shared problems. This is an excellent time to explore the interface of citizen science with new online tools that allow for direct social interaction among participants.

Until the early twenty-first century, participants in Internet-supported citizen science projects, such as the large projects delivered by the Cornell Lab of Ornithology, acted alone or in small groups with little direct interaction apart from Listservs. As projects continue to use more personalized Internet-based social media, including social networking tools, citizen scientists are likely to become better "connected." What does current understanding of social networks tell us about the potential impact of combining social networking with citizen science?

One body of theory that may help to increase understanding of the potential impacts of citizen science is collective action theory. Citizen science attempts to harness joint participation of professional scientists and nonprofessionals for the public good. The public good outcomes of citizen science include data collected by participants, which can be used to study shared environmental problems occurring over large spatial and temporal scales, and the knowledge produced when citizens and professionals use the data. In this chapter, we characterize citizen science projects as collective actions and consider how collective action and social networking theory might improve project designs to extend impacts beyond the traditional program goals to include the public good outcomes of scientific discovery, public knowledge, and, ultimately, improved policy and management.

The goals of citizen science projects featured in this book are to improve nonprofessional participants' knowledge, enable communities of practice interested in science and the environment, promote scientific discovery, and improve environmental stewardship. To the extent that public knowledge and improvements to environmental quality are shared by all, citizen science can be considered a collective action.

Large-scale citizen science initiatives (e.g., continent-wide projects like FeederWatch) have begun to allow participant interactions to occur both synchronously and asynchronously through Internet forums and social network sites (e.g., Facebook), and the Cornell Lab's newest project, YardMap (www.yardmap.org), has its own integrated social network that can push information out to Facebook and Twitter. Although Internet-based social networking tools are not yet fully integrated into most projects, we believe that such integration will change how participants interact with each other over project-related activities and create online communities of practice (Wenger 1998). Increased participant interactions may enhance social learning among participants (Merriam et al. 2006), which in turn could enhance the quality and quantity of data collected and the public knowledge outcomes.

Citizen Science as a Collective Action

In citizen science, the "partnership between volunteers and scientists to answer real-world questions" (Cohn 2008) is an example of a collective action. Collective actions occur when two or more people pursue common goals to achieve outcomes that require group effort (Olson 1965). A key feature of collective actions is the public goods outcome, which is freely shared (nonexcludable) and nonrivalrous (one individual's use does not diminish another's use). Here we consider citizen science to be a collective action in a broad sense, wherein a variety of contributors invest in the collective by providing project support, teaching people how to collect data, collecting data, helping get the word out, conducting research with the data, using the data to educate others, making observations using dynamic mapping and visualization tools, and writing scientific papers or popular articles based on the data. The basic outcomes are data, knowledge, and discovery.

To the extent that citizen science provides information or supports pol-icy or management activities that improve conservation efforts (Rosenberg, Hames, et al. 1999; Rosenberg et al. 2003), collective citizen science can also be considered a public good—one that is sharable by all whether or not they participate in citizen science.

Many opportunities exist for people to contribute to the collective citi-zen science effort. Large, Internet-based citizen science projects ask peo-ple to participate in collecting data and enter them online where they are shared with scientists and participants. Professional scientists are interested in the data for academic research or management, whereas the participant community may have additional interests, such as learning about patterns and trends, using the data in classroom exercises, or understanding how their data compare with those of others nearby. In citizen science projects that involve direct action, such as planting native trees or bird gardens and studying their impacts, the shared resource may be improved habitat qual-ity and even the trees or birds themselves (e.g., Cooper et al. 2007).

The concept of participants working with scientists to conduct research of vital importance to understanding landscape change, or working to study how landowners might achieve larger (collective) objectives that require widespread participation, exemplifies the collective action concept (Ostrom 1998). While citizen science monitoring initiatives emphasize exploring and managing biodiversity on privately owned lands (Cooper et al. 2007; Sinclair and Knuth 2000; Wagner et al. 2007), the idea of private citizens working together for larger (collective) objectives is not new. Landowner cooperatives or wildlife management associations have existed for decades, but they are found mostly in rural landscapes and commonly target habitat management for special interests, such as people hunting deer and other game (Wagner et al. 2007). When individuals engage in sustained cooperation for the collec-tive good, the collective action is successful, averting tragedy of the commons, where common-pool resources are prone to degradation (Hardin 1968; Wag-ner et al. 2007). A benefit of bringing together collective action with social network theory is that it opens up the possibility of discovering new ways to avert the tragedy of common-pool resources. Models have shown that collec-tive actions can be sustained by social interactions (e.g., social affirmation, reputation, or punishment) even when interactants are anonymous (Hauert et al. 2002; Mathew and Boyd 2009). Because online social networking tools have begun to create a social environment where individuals' contributions are visible to others, individuals may establish identities and reputations that help sustain collective action within the context of citizen science.

Social and Psychological Considerations in Collective Action

In some cases, individuals choose to participate in collective actions be-cause they form a group identity and perceive the efficacy of the impact of collective actions on the group (Klandermans 2002). There is also a wealth of evidence indicating that group identity and social diversity have significant subconscious components and motivations that may influence participation in collective actions (cooperation) to improve environmental quality (Santos et al. 2008; Dickinson 2009). Still others (Ostrom 1998; Wagner 2007) suggest that norms, identities, and trust are important factors in predicting whether people participate in collective actions. The interface between these bodies of evidence and collective action theory is a fruitful area for future research.

Moving beyond simple data collection, an Internet-based citizen science project can enable participants to form self-governing institutions to regulate individual behaviors related to citizen science. Self-governance could be enhanced with integration of social networking, which allows people to communicate with each other, potentially fostering shared interest in the outcomes. Over time, social networking might offer participants the opportunity to develop a group identity, establish norms that members are "expected" to adhere to, consider sanctions, and trust each other's contributions toward cooperative environmental management.

How People Participate in Citizen Science Collective Actions

Citizen science participants (e.g., volunteers, scientists, and observers) may work toward common goals by contributing data, engaging in discussion of the data, writing papers based on the data, translating the results for a broad public, translating the results into management recommendations, and carrying out management recommendations on private or public lands. Some individuals who are not personally involved may nevertheless benefit from the outcomes of others' actions and can be considered free riders. An example is people who benefit from the conservation outcomes of research and management activities without contributing data themselves.

We offer a second example of free riding within the context of eBird, an online birding checklist project, and a new iPhone application, Birds-Eye, which was developed to help the birding community take advantage of eBird data to locate interesting birds. One can imagine free riding with BirdsEye, because the app uses the eBird database to point people to sites where particular birds have been seen. It is like a handheld rare-bird alert system—birders who land at an airport can immediately look at what birds have been seen that day and take off in search of a "life bird." Those who use the iPhone application without joining or contributing data to eBird are effectively free riding because they are using the eBird database to find birds they wish to see without contributing their own observations to the database so that others will be able to do the same. Still other individuals who do not use BirdsEye may be considered free riders if the data from eBird are used to inform conservation or management actions. As this example shows, understanding why people participate in Internet-based environmental collective actions (i.e., do not free ride) may be critical to expanding involvement in citizen science initiatives.

Possible Effects of Enhancing the Number of Connections across Large Distances

If collective actions depend on face-to-face training and communication, the scale will be limited by people's ability to establish and maintain relationships with others who are interested in the same goals. It remains to be seen whether Internet-based social networking will help overcome the geographic limitations on face-to-face communication by enabling citizen science participants to establish and maintain relationships with scientists and other participants across vast distances, enlarging each participant's social network beyond his or her day-to-day world (Figure 15.1). Understanding how these connections increase the potential for cooperation is vital to understanding how citizen science, by virtue of integrating social networking, can increase its scope and scale of impact.

Influence of Leadership and other Social Factors on Diffusion of Cooperation in Social Networks

An individual's decision to participate in a collective action is optional and the outcomes of joint participation are available to the broader community,

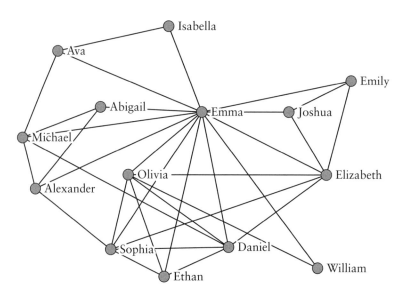

Figure 15.1. Example of a citizen science social network graph.

including the free riders (Rogers 2003). When deciding to cooperate, people often rely on opinion leaders or other members of their social network for social affirmation regarding whether to contribute to collective actions for public goods (Rogers 2003; Strang and Soule 1998). To understand how citizen science might enjoy increased participation (reduce free riding), we explore how beliefs or behaviors (i.e., innovations or collective actions) spread through a society via a diffusion process.

Individuals go through five major stages when determining whether to accept or reject an innovation (i.e., a behavior). Those stages are (1) *knowledge* of a behavior; (2) *persuasion* by others to adopt or reject a behavior; (3) *decision* to adopt or reject a behavior; (4) *implementation* (i.e., actually doing the behavior); and (5) *confirmation* of the decision to adopt or reject a behavior. Acquaintances or opinion leaders in one's social network provide subjective evaluations (cues) about whether to engage in a specific behavior.

How Do We Study the Impacts of Social Networks?

Social network research investigates the importance of relationships among individuals, occurring in person, online, or in both realms, where the relationship is the unit of analysis (Scott 2000). Social network theory and analysis allow researchers to examine the classic questions of how social relationships influence people's learning, thinking, and behavior (Granovetter 1985) and are appropriate for understanding why people participate in citizen science. Social networks via the Internet may increase participation in citizen science by creating communities of practice that increase awareness of others' participation, quickly spread new ideas, and allow individuals to create or maintain relationships that influence each other's knowledge and behaviors. We explore three theoretical domains in social network research and relate them to citizen science.

Contagion and Weak Ties among Individuals

Social contagion is the spread of new information and ideas through society that occurs as a consequence of people interacting with each other (Burt 1987). Contagion theories have provided limited explanation for why people adopt behaviors in collective action (Centola and Macy 2007; Diani and McAdam 2003), but they do explain how people become aware of new ideas or behaviors. Such awareness can spread through a group when people are minimally socially involved with each other—that is, when they have weak ties. The "strength of weak ties" theory describes the impact of loose acquaintances when one acts as a local bridge for information transfer by connecting two people who are dissimilar (Granovetter 1983; Rogers 2003). Weak ties are good for diffusing information and providing access to new information, which is critical for the first stage in which potential participants are made aware of opportunities to participate in citizen science and other collective actions (Rogers 2003). It is possible that Facebook connections through relatively few "friends" are strong ties, and "friends of friends" represent many more weak ties.

Strong Ties among Individuals

In contrast, strong, affective, and time-honored relationships, presenting as strong ties among individuals, are useful during times of uncertainty and change (Krackhardt 1992). The social network concept of strong ties is critical for persuading others to act (i.e., adopt a behavior) and providing affirmative feedback after the behavior has been implemented. When considering adopting new behaviors, individuals tend to rely on social affirmation from those with whom they have strong ties (Centola and Macy 2007). Citizen science participation may spread more rapidly among individuals who have strong ties, and strong ties may also facilitate behavioral change (Rogers 2003).

Does Like Attract Like? Homophily in Social Networks

The proverbial expression "Birds of a feather flock together" exemplifies the social network pattern of homophily, where individuals have higher rates of contact with people who have similar characteristics than with people who have dissimilar characteristics (McPherson et al. 2001). Networks may cluster by *status homophily* of race, gender, social status, age, education, or occupation, or by *value homophily* such as values, attitudes, or beliefs, which are presumed to shape orientations toward future behaviors (McPherson et al. 2001). Citizen science social networks may be homophilous on a variety of dimensions—for example, knowledge—and these homophilies may be caused by geographic proximity, conservation interest, or even taxonomic preference. Internet-based social networks allow for collection of digital data on how networks coevolve over time, including how value homophily enables or constrains the behavioral outcomes. Homophilous network relationships might limit one's knowledge of opportunities for participation, which may run counter to the goals of expanding information flow among citizen science participants.

Social Media and Social Networking

The Internet allows collaboration, entertainment, and sharing of opinions, art, videos, or images (Wikipedia 2009). Nearly half of all Americans

(48%) subscribe to social media accounts such as Facebook or Myspace (Harris Poll 2009), a figure that suggests strong interest in building or maintaining relationships with others via the Internet. Research on computer-mediated interaction suggests that although people are able to build meaningful relationships with each other over the Internet, it takes more time than face-to-face relationship building does (Boyd and Elison 2007; Walther and Burgoon 1992). Internet-based social media offers a new lens through which to look at how people participate in citizen science and how to knit biological monitoring together with collective actions that directly influence conservation. It may be possible to expand citizen science to broad spatial (e.g., continental) scales more quickly with social networking. Social networks lead to more frequent and timely flow of information among participants, which may help coordinate collective actions (Weare et al. 2007). The Internet also makes it easier for individuals who might not otherwise become involved, because it is easier to participate in collective actions when they are largely communicated and conducted online (Postmes and Brunsting 2002). Weak tie relationships may facilitate the spread (contagion) or awareness of new information, whereas strong tie relationships may facilitate the adoption of behaviors—for example, collecting data or managing one's landscape in a particular way.

On the other hand, if the overall network structure of a project is homophilous, network members may not receive new information or social affirmation required for diffusion of innovation (new ideas or practices) or adoption of behaviors because they are interacting with individuals who, for the most part, are just like themselves. This hypothesis depends on the overall degree of similarity of behaviors and ideas within a group, however, and it is possible that an intermediate degree of homophily allows sufficient differences in information content and behaviors so that diffusion will not be significantly impeded. Further, homophilous groups, although not large, may be very important as centers of expertise and high-quality data collection.

How might tie strength play into this equation? Experiments have shown that small similarities, such as birth date and first name, influence the likelihood of friendship, compliance, and even behavioral mimicry offline, suggesting that it may not take much in the way of similarity to create new affiliations online (Burger et al. 2004; Guegan and Martin 2009). Social affirmation through friendships may increase sharing of new information and influence as well as retention rates and extent of involvement of participants in projects. The interactions between homophily, tie strength, and diffusion/contagion are complex and worthy of additional study.

Individuals form relationships with each other on social media sites, convening over meaningful experiences or activities, such as supporting a cause, discussing the history of their hometown, or connecting with others through their alma mater (e.g., Facebook groups). Forming relationships through engagement with citizen science should be no different as citizens can provide meaningful information about their conservation experiences or activities (e.g., monitoring birds or reporting landscaping practices). As shared meaning implies attitude similarity, citizen science participation may create conditions favorable to kinlike cooperation, in which similarity of meaning behaves as a heuristic cue for kinship among people who are not related (Park and Schaller 2009). If so, the level of cooperation would be much higher than generally expected from collective action theory.

Conservation-based Internet social networks may enhance face-to-face relationships, but coordinators should be careful not to reinforce homophilous networks in a way that creates a divide between those who engage in citizen science and those who do not. Ideally, program coordinators should encourage citizen scientists to invite people with whom they very rarely interact to join the network, creating an opportunity for information exchange and social affirmation for conservation-related collective actions. In practical terms, this constitutes moving away from the "preaching to the choir" approach that characterizes many citizen science projects. On the other hand, there is some concern over the potential development of ideological rifts between different groups of participants within the same project. This sort of conflict has been known to occur on Listservs, such as Bluebird-L, which was retired in 2009. A frustration with that Listserv was that arguments would break out periodically over the pros and cons of destroying nests of the invasive House Sparrow (*Passer domesticus*). These would sometimes get out of hand, and certain individuals were removed from the Listserv for inappropriate dialogue. Balancing the costs and benefits of homophily is a challenge that is poorly understood, warranting further study before predictions can be made that are informative to citizen science program designs.

How Social Networking Can Be Expected to Enhance Educational Outcomes

Internet-based social networking may enhance learning outcomes in citizen science by allowing opportunities for participants interested in conservation to socially interact and influence each other. The process of interacting with and observing others in a social context is social learning, which is manifested by socialization and a belief that behavior is controlled by the individual learner (Merriam et al. 2006). A group of adults with common goals who engage in social learning as they seek to achieve those goals is considered a community of practice (Wenger 1998). Online communities of practice enable adults to interact together, share information, master material, and become innovative. Social networks also influence educational outcomes when adult learning is viewed through the behaviorist lens, whereby learning is achieved when individuals exhibit specific desired changes in knowledge, attitudes, or behaviors, including achieving competence in a subject and acting on that competence (Merriam et al. 2006). We have three stages of learning outcomes desired for citizen science participants: (1) *basic*—data collection skills, ability to follow protocols, and correct identification of organisms; (2) *intermediate*—conceptual understanding of questions and their relation to results; and (3) *advanced*—inquiry, problem solving, and invention. Individual learning outcomes can spread through friendship networks (Mason et al. 2008) or be channeled from "opinion leaders" to followers (Valente 1996). We still do not know whether social learning through the Internet shares the benefits observed in other interpersonal contexts, but if it does, the number of people affected will far surpass what is typical for local, in-person projects.

Future Research

An Internet-based social network, if integrated with a citizen science program, might fundamentally change the way participants interact with others in the program. Current programs provide little opportunity for interaction and typically do not provide information about participants. As such, participants do not know who other participants are unless they are featured on the website for making a special contribution or winning a prize. Because programs are delivered over the Internet, participants do leave a digital footprint of their observations and activities for potential longitudinal analyses. When social networking is added, researchers will be able to observe how participants form social networks and interact with each other and thus determine how network interactions influence individual behaviors, collective actions, and behavioral outcomes. Researchers can move beyond evaluation of learning impacts to design studies for multilevel analysis of individual characteristics, network relationships, network structures, and their influences on the potential for collective action (i.e., contributing information to the collective knowledge base or implementing management actions for the public good). These research possibilities are discussed further below.

Call for a Merging of Research Areas

How much does participation in citizen science influence individuals' attitudes, beliefs, or behaviors, and how much do individual characteristics explain decisions to participate in citizen science data collection or management practices? These are difficult questions, especially because individuals' psyches or identities may play a strong role in motivating participation in a program. What characteristics do opinion leaders possess, and do citizen science participants exhibit deception to maintain their leadership status? Is identity developed through program participation, and does social networking facilitate this? All of these questions will inform our understanding of the cognitive and psychological underpinnings of citizen science participation, social network interactions, and the wide array of outcomes that citizen science projects hope to achieve.

Social network relationships potentially influence access to information, how people feel about the information they receive, and their reactions to the information in the form of attitudes and behavior. But the mechanics of social networks also influence these outcomes, suggesting that it is only by understanding how social networks function that we can predict the outcomes of participation in social networks tied to citizen science.

Understanding the characteristics of social networks is an essential component of understanding the relationship between social networking and citizen science participation. How do cliques form? Are there key gatekeepers brokering or constraining information flow? To what extent is the network structure homophilous or heterophilous, and how does this enable or constrain social learning and network influence on adoption of information or behaviors? Collective action behaviors sometimes vary according to network position (Oliver et al. 1985). For example, core nodes may engage in all behaviors, whereas peripheral nodes may engage only in a specific subset of behaviors. It is not a given that social networking will augment all of the impacts that citizen science is trying to achieve; impacts that are ideologically based may be more resistant to change than behaviors that are less value laden. Further integration of the quantitative aspects of social network analysis with the social and psychological attributes of individuals is critical to understanding more fully the impact of social networks on citizen science outcomes.

Because the goal of many Internet-based citizen science programs is to incorporate information spanning broad spatial (continental) and temporal scales, the opportunity exists to analyze nested data at local, regional, or national scales. Additionally, because the geographic locations of the data and users are known, investigators can determine the influence of spatial relationship on behaviors for comparison with the influence of social networks (regardless of proximity) on behaviors.

Researchers can also design longitudinal studies that track socially networked participants over multiple points in time. Because individuals create user profiles to interact with each other in social networks, researchers can easily contact the participants for self-administered questionnaires or track online behaviors. Some research questions that might be addressed are, How do participants change over time? How do networks evolve? How do intended (predicted) behaviors align with observed behaviors? What factors influence whether people continue to participate?

✦

We are seeing tremendous potential for research in this relatively unknown arena with significant promise for improving the learning, scientific, and behavioral outcomes of citizen science programs. What type of feedback and incentives are needed to continue motivating individuals to participate in citizen science programs? Do individuals engage in citizen science monitoring and pro-environmental behaviors because of their own decisions, or are they influenced by their social networks (i.e., social affirmation), perception of others' behaviors (i.e., the knowledge that others engage in collective actions), or by actually seeing the impacts of pro-environmental behaviors? By building an infrastructure to support social sciences research as new citizen science programs are developed, practitioners can set the stage for investigation of a diversity of questions about the nature of human cooperation. Projects created with this knowledge could be a major force in supporting buy-in, understanding, attitudinal and behavioral change, and connections required to solve environmental problems whose cause is distributed across vast space.

Acknowledgments
This research was supported in part by the Cornell University Agricultural Experiment Station federal formula funds, project no. 147433, received from Cooperative State Research, Education, and Extension Service, U.S. Department of Agriculture, and by a grant from the NSF Informal Science Education Program (DRL-ISE#0917487) to Janis L. Dickinson. Any opinions, findings, conclusions, or recommendations expressed in this publication are those of the authors and do not necessarily reflect the view of the U.S. Department of Agriculture. We thank our colleagues in the Human Dimensions Research Unit for their thoughtful critiques of this manuscript.

16

A Role for Citizen Science in Disaster and Conflict Recovery and Resilience

KEITH G. TIDBALL AND MARIANNE E. KRASNY

How might we integrate concepts from citizen science with scholarship and practice aimed at fostering community capacity to buffer the impacts of disaster? Drawing from the disaster, conflict, natural resource management, and resilience literatures, and from examples of participatory data collection linked with environmental restoration in postcrisis settings, we hope to stimulate thinking about how commencing citizen science activities both prior to and after large-scale crises might contribute to social-ecological system resilience in situations of human vulnerability.

Weinstein and Tidball (2007) argue that the deficit-based approach to disaster response, an approach that focuses on absences and weaknesses, is inherently constrained to actions that ignore both the importance of building or repairing feedback loops and the change agents that make these loops relevant. In fact, they argue, the resulting "imposition" of aid following deficit-based needs assessments often mobilizes change agents against seemingly sensible inputs by the "benevolent developer" (ibid.; United Nations and World Bank 2004). Further, according to stakeholders in Africa quoted in the New Partnership for Africa's Development (NEPAD) (2005) document, "externally driven post-conflict reconstruction processes that lack sufficient local ownership and participation are unsustainable. They cause resentment and fail to integrate the underlying socio-cultural belief systems that shape the worldview of the internal actors" (New Partnership for Africa's Development 2005:20). The implications of these concerns have resulted in the initiation of two important shifts in thinking in disaster and conflict response contexts. The first is that asset-based participation among affected populations, focused on strengths, opportunities, and assets rather than exclusively on deficits, is required to identify acceptable or desirable assistance. The second is the acknowledgment of the necessity to account for (usually perception-driven) self-reinforcing growth trends, or positive feedback loops.

We propose that citizen science is poised to contribute positively to the above-mentioned shifts in thinking. We forward the notion that citizen science can build capacity to mitigate disaster and conflict in three important ways: first, through facilitating local knowledge creation, ownership, and participation; second, by initiating and reinforcing desirable feedbacks; and third, by amassing multiple forms of knowledge and data collection over broad geographic areas. Further, considering the need for asset-based and participatory interventions during and after a conflict, and acknowledging the paucity of mechanisms that currently address this need (Weinstein and Tidball 2007), we feel that citizen science should be examined by the crisis response community for its potential to become part of a tool kit of participatory responses that engage citizens in meaningful postcrisis activities.

Disaster and Conflict

For the purpose of this chapter we use a broad and generally accepted definition of disaster, which is a situation combining a potentially destructive agent or force with a population in a condition of vulnerability, resulting in disruption of individual and social needs for survival, social order, and meaning (Hoffman and Oliver-Smith 2002). We include natural processes and processes that result from human intentionality such as conflict or warfare—for example, recent intervention and conflict in Iraq.

The frequency of disasters has increased worldwide over the period from 1989 to 2008 (Rodriguez et al. 2009:33). In this same period, however, human adaptation and the adoption of new infrastructures to mitigate disaster, including detection and warning systems, have reduced disaster deaths by creating programs for community-based disaster preparedness, evacuation, and mitigation (IFRC 2009b; Rodriguez et al. 2009:33). Thus, the capacity of a community or nation to reduce the potential negative consequences of risk, together with its vulnerability and the nature of the hazard itself, define whether a calamitous event becomes a disaster (IFRC 2009b).

Although natural disasters may be attributed to single and immediate events, such as hurricanes, the importance of social, ecological, and political factors in determining both the likelihood of disasters and ability to respond to them is increasingly apparent. For example, the catastrophic events following Hurricane Katrina were linked not only to the devastating storm but also to ongoing building and ecological degradation along the Gulf Coast, class and ethnic differences, governance issues such as institutional capacity and leadership, and possibly global climate change (Ernstson et al. 2010).

The causes and impacts of conflict are even more complex than those of disaster at multiple levels including ecological, social, political, economic, and regional. An example of both nation-state and intergroup conflict can be seen in the wars in Iraq and Afghanistan, where initial outside intervention by the United States and other countries failed to ameliorate and perhaps contributed to destabilization, a situation that some argue led to increases in violence between ethnic and religious groups within those countries (Ferguson 2006). Yet government responses to conflict situations often are characterized by lack of recognition of multiple interconnected impacts. They discount local knowledge, cultural norms, and ongoing local self-organizing practices that could help mitigate conflict (Weinstein and Tidball 2007). This neglect can lead to externally imposed, on-the-ground interventions that work against larger strategic goals (Tidball et al. 2008; Weinstein and Tidball 2007).

Self-Organized and Participatory Responses to Disaster and Conflict

Within the disaster response communities, concern about ineffective outside responses has caused debate about the role of local people in developing capacity to prepare for disaster or conflict and to respond after calamity has struck. For example, in reviewing the recommendations of an international conference on community risk assessment, Pelling (2007) noted that participatory disaster risk assessment can build local capacity and generate knowledge that, along with more expert-driven data collection, can help identify and reduce the risk of disaster. Similarly, Weinstein and Tidball (2007) and Tidball et al. (2008) presented an alternative model for post-conflict intervention, "civic ecology practice," which is based on local assets (including self-organized natural resource management practices) such as community forestry, watershed enhancement, community agriculture, and gardening (Tidball and Krasny 2007). Though citizen science has not yet been included by the postcrisis management community in the "menu" of approaches that help communities manage their natural resources postcrisis, the similarity of citizen science to civic ecology practices suggests that a citizen science approach could help mitigate postdisaster and postcrisis situations.

Civic ecology practices emerge through the actions of local people wanting to manage a particular resource. They integrate learning through small-scale experimentation and observations (adaptive management) with collaborative or participatory processes (comanagement), and thus might be considered an emergent form of adaptive comanagement (Armitage et al. 2009; Krasny and Tidball 2009; Ruitenbeek and Cartier 2001). Civic ecology practices share important attributes with citizen science, as both can be considered "public participation in scientific research" as described in Bonney, Ballard, et al. 2009. The local knowledge of individuals who initiate the practice is critical, although linkages are often made with scientists from universities, government, and nonprofit organizations and thus multiple forms of knowledge are incorporated into the stewardship activities. Further, social and adaptive learning occur as participants try out and assess the results of different restoration practices and ways of engaging their broader community, sometimes becoming increasingly involved in the political process when the lands they manage are threatened by outside interests (Tidball and Krasny 2007). This learning shortens feedback times between management actions and the impact of those actions, such as adopting participatory approaches for planting trees and seeing the impact of tree planting on local ecological and social systems. Thus, civic ecology embodies a number of attributes that characterize social-ecological systems that are resilient both prior to and after a crisis, including multiple forms of knowledge and governance, self-organization, adaptive learning, shorter feedbacks, and ecosystem services (Folke et al. 2002; Walker and Salt 2006).

What Role for Citizen Science in Disaster and Conflict?

We propose that citizen science could build capacity to mitigate disaster and conflict through reinforcing desirable social and environmental trends and feedbacks, and by bringing to the table multiple forms of knowledge and data collection over broad geographic areas. Further, given the need for asset-based and participatory interventions during and after a conflict, as described above, citizen science should be considered for its potential to elicit participatory responses that engage citizens in meaningful and empowering postcrisis activities.

Fundamental to discussions about civic ecology, citizen science, and community-based approaches to mitigating the effects of disaster and conflict are issues related to participatory means of engagement. There is growing support for the notion that when citizens are engaged in collecting data and in related social learning they will build their own capacity to adaptively manage risk and local environmental resources (Blackmore 2007; Bonney, Ballard, et al. 2009; Bonney, Cooper, et al. 2009; Pahl-Wostl et al. 2007; Pelling 2007). Therefore, the potential for this kind of engagement to be applied to build adaptive capacity in postdisaster contexts seems like a logical next step. Thus, in this section, we briefly describe examples of citizen data collection in postcrisis settings. Although these examples are not directly analogous to citizen science as a distributed means of data collection, we hope they will stimulate thinking about how bringing knowledge from the social sciences to the development of citizen science programs might allow citizen science to play a larger role in building community capacity to prepare for and respond to disaster and conflict.

Identifying Pollution Hot Spots

In December 2008, a massive spill of toxic coal ash fanned out across 300 acres of countryside and waterways in Tennessee. Whereas Tennessee Valley Authority (TVA) officials downplayed the spill, claiming that the material "does have some heavy metals within it, but it's not toxic or anything," scientists and environmental groups released more worrisome findings, including elevated levels of arsenic, barium, cadmium, lead, and selenium in water at the spill site (Newsweek 2009). According to Bill Kovarik, a scientist involved with the nonprofit group that responded to the spill,

> the TVA coal fly ash disaster marked a turning point for environmental reporting and environmental activism. For the first time, the media had sampling results from environmental organizations almost right away. Robert F. Kennedy Jr.'s Riverkeepers, especially, were very active in gathering samples and getting them to university and government toxicology labs. (Dawson 2009)

Kovarik goes on to note that in contrast to usual processes of public concern, followed by government agency denial and then ebbing public interest,

> the typical cycle was broken wide open in the TVA disaster. News organizations were in the air taking photos, and environmental groups were in the water taking samples, within hours of the disaster. . . . TVA said they found arsenic in the river at 40 times below the drinking water standard, but the environmental groups found arsenic levels 300 times higher than the drinking water standard. (Ibid.)

These quotes suggest that through the data-collection efforts of citizens who had become voluntary affiliates of local environmental groups, the TVA disaster was better documented, leading to greater government accountability.

Assessing Risk and Damage

Pelling (2007) describes several participatory approaches to risk assessment and postdisaster response that could be expanded to include citizen data collection. In one scenario, key themes related to disaster readiness and response are identified through participatory assessment tools such as focus groups and key informant interviews followed by more quantitative methods such as structured surveys. Examples include a food security and rural livelihoods survey coordinated by the USAID Famine Early Warning Systems Network and an effort by Save the Children to understand how poor farmers coped with the 1992 drought in southern Africa (Pelling 2007). In an example from Jamaica, the Water Authority had trouble obtaining data on water levels from remote sites and enlisted volunteers to read river gauges. These individuals moved to assigned locations where they reported readings of high water levels, thus providing an opportunity for protective measures to be implemented prior to flooding (IFRC 2009a:210). In another instance the Gulf Coast Tree Assessment, a collaboration of state agencies, nonprofit organizations, and volunteers, assessed the condition of more than 7000 trees in ten separate Gulf Coast communities following Hurricane Katrina. One of the future goals of the project is to establish a predisaster Urban and Community Forestry Databank (Hartel 2006). We acknowledge that in all of these cases, what is traditionally considered citizen science is not clearly present. We argue, however, that it could be. Should citizen science be employed in these cases, both the participatory data collection and resulting data, as well as the social benefits of local involvement, feelings of self-reliance, and local ownership, would all enrich and enhance disaster recovery efforts.

Restoration of Disturbed Social-Ecological Systems

In southern Iraq, the Mesopotamian Marshes once covered 9000 square kilometers and harbored the Marsh Arabs, who derived their food and fiber needs from the marsh ecosystems. In retaliation against Marsh Arab activism after the first Gulf War, Saddam Hussein built a series of dikes to divert the water, resulting in an environmental and social disaster and forcing the displacement of marsh residents. After the fall of Hussein in 2003, some Marsh Arabs returned to the site, breaching the dikes in an attempt to reestablish the wetland ecosystem and their traditional way of life. Aided by USAID and other development organizations and scientists, Iraqi biologists have enlisted Marsh Arabs to help them take ecological measures of marsh restoration and thus participate in research to restore the wetland systems (Alwash 2005). In another instance of citizen participation in natural resources science and management after a conflict, the NGO Bird Life cooperated with Iraqi biologists to conduct winter surveys of Key Biodiversity Areas in sixty-five sites across Iraq (Birdlife International 2009). What makes these examples different from typical citizen science projects focused on birds is that their initiation is part of wider efforts to recover from large-scale disaster or war. In these and other cases, the benefits of doing citizen science are measured by citizens in terms of enhancing the likelihood of recovery of their lives and landscapes.

Adding Local Data to Global Databases

According to Pelling (2007), rapid progress has been made in the development of global indices of national vulnerability and disaster risk, such as the United Nations Development Programme's (UNDP) Disaster Risk Index and the World Bank's Hotspots project. Participatory methods of disaster risk assessment could complement these global indices to provide more locally grounded assessments of vulnerability. Through the process of aggregating data from local efforts into global assessments, local actors can participate in deliberative processes such as comparing experiences, measures, causes, and responses to risk and vulnerability across multiple sites. Further, Pelling (2007) suggested that "the opportunities for strengthening local governance and bringing communities together, especially in unregulated urban districts or among populations suffering from social division, might be more useful than the data outputs derived, although these would be required to provide a framework for inter-communal discussion and capacity building" (Pelling 2007:383). For example, though tree damage assessments in New Orleans after Hurricane Katrina were of marginal scientific use, the value of these efforts—which closely resemble citizen science (and could easily *become* citizen science)—to the bolstering of community solidarity and strengthening of local governments cannot be overstated (Tidball and Krasny 2008b).

❦

Citizen science traditionally has focused on balancing research and educational goals through citizen and student data collection in large-scale monitoring and research projects (Krasny and Bonney 2005). In so doing, citizen science has stimulated thinking and programs that address issues more generally of public participation in scientific research (Bonney, Ballard, et al. 2009).

In addition to citizen science, other fields of endeavor, including natural resources management and disaster and conflict preparedness and mitigation, have similarly addressed issues of participation, such as balancing the role of expert and local knowledge and engaging community members in data collection. We have provided examples to make the case that there is room, perhaps even a need, for citizen science approaches within the disaster and conflict relief and mitigation communities.

Berkes and Folke (1998) speak to the challenges of finding suitable social mechanisms that help confer resilience on social-ecological systems. They further suggest that, in order to maintain function in the face of disturbance, such systems need to be able to recognize feedback, and therefore require *"mechanisms* by which information from the environment can be received, processed, and interpreted" (ibid.:21–22). We have shown how citizen science, through becoming integrated into asset-based and participatory approaches to risk assessment and disaster response, and by fostering learning among participants in adaptive comanagement, might become one social mechanism that can shorten feedbacks that inform stakeholders about the effectiveness of their management actions. Further, we have explored how citizen science might become part of efforts to transform undesirable feedback loops commonly found in postconflict and postdisaster settings, in which violence and loss of infrastructure lead to lack of meaningful employment and educational opportunities, thus creating further violence and degradation of infrastructure and the environment. As an alternative, efforts in which local people collaborate in self-organized resource management initiatives, leading to social connectedness and enhanced ecosystem services, may create conditions that foster learning and the ability to engage in meaningful livelihoods (Tidball and Krasny 2008a). Engaging citizens in data collection in conjunction with their other stewardship activities has great potential for expanding opportunities for learning and adaptive management in such postdisaster and postconflict scenarios.

Acknowledgments
The authors would like to acknowledge helpful comments and suggestions to this chapter by the book editors. We would also like to acknowledge critical insights into postconflict and postdisaster contexts from Elon Weinstein, and for helpful direction from Assam Alwash.

Afterword

JOHN W. FITZPATRICK

Perhaps our greatest gift as humans is that by virtue of advanced cognition, imagination, and capacity for observational learning, we are born to be curious animals. Beginning in our earliest years of life we explore our physical and social worlds with powerful internal drives to discover and name things that surround us, to recognize relationships among them, and to make higher-order sense out of these relationships—including where we ourselves fit in. We cannot help wondering how and why these things and relationships exist. Science was born out of these innate drives. In its elemental sense, science is nothing more than an organized, collective expression of human curiosity about the world around us.

In this context, every human in the world is born a scientist. It is a sad fact of twenty-first-century global society that most individuals are also born into social, political, economic, or religious worlds that divert or—worse—suppress our ability to practice our innate proclivities for acting like scientists. Even in societies that do value and foster individual curiosity, educational systems have been slow to adopt the notion that training in science should be built around exercising the gift, as opposed to memorizing facts and formulas. Happily, the advent of inquiry-based science curricula has begun to correct this tendency, and it is no coincidence that citizen science is playing an important role in this emerging emphasis on individual inquiry (see Chapter 12).

As described and explored throughout this book, citizen science represents an extraordinary new stage in democratizing the organization of human curiosity. A burgeoning array of scientific projects, most having Internet-enabled entry points, makes it possible today for virtually anyone to take part in the scientific process, and in so doing to practice being curious. Certainly, the explosion of local, regional, and globally scaled citizen science projects belongs among the most profoundly important social consequences of the development of the Internet (see Chapter 15).

It is important to temper our enthusiasm somewhat by acknowledging a few obvious limits. Not every field of inquiry or type of question is amenable to being addressed through direct participation by the general public. Particle physics, genome sequencing, nanoengineering, and hundreds of other lines of scientific inquiry will remain largely out of reach for untrained individuals. Even in fields like ecology and natural history, where citizen science is developing rapidly, most of the key questions still require specialized knowledge, expensive instrumentation, intensive longitudinal studies of marked individuals, quantitative statistical methodologies, and other specialties that demand lots of money, time, advanced training, or all three.

As this book demonstrates, however, the arenas in which masses of citizens can practice and contribute meaningfully to scientific progress are expanding rapidly. The best of these projects are grounded in two simple truths about curiosity: (1) the cornerstone of curiosity is observation; and (2) when curiosity is rewarded with a return of knowledge, it is not satiated, but instead grows deeper and more committed.

Observation: The Fundamental Unit of Curiosity

The most successful applications in which citizens are helping ask and answer scientific questions are designed to treat each individual as a "sensor." By encouraging people to record and report some piece of information about a known place at a known time, citizen science projects convert the fundamental unit of individual curiosity—simple observation—into data. As it happens, not only are many ecological and environmental questions data hungry but the data they require can be quite accessible to the general public. Hence, as this book confirms, environmental sciences have taken the lead in developing the approaches and technologies required of citizen science (see Chapters 1 and 6). Dozens of environmental projects are now organized to answer genuine questions posed by scientists, delivered via Internet portals created by Java programmers and Web designers, and analyzed by statisticians and geographers. In this form of science, the key challenge for all these advanced specialists is to provide motivating contexts that induce comparatively untrained individuals to look out their window, or step outside, or take a walk, or visit a stream, and—most important—to record what they see and submit the records, thereby converting their observations into data. Adding simple protocols to this exercise of observing adds enormous additional power to the capacity for citizens to help answer specific scientific questions.

As the field of ecology matured throughout the twentieth century, the dynamic nature of the natural world emerged among its central tenets. Today, most of the important questions in ecology require information about change across both space and time. As emphasized throughout this book, such questions frequently seek data at spatial and temporal scales that are extremely difficult to measure by conventional scientific approaches, or during conventional funding cycles. Accurate distribution maps of species at fine geographic scales across space, at different times of the year, and among years, require literally thousands of data points gathered across whole continents. The new approaches required to address the enormous volumes of data produced by these networks of citizen sensors across the globe have pushed researchers into uncharted territory in developing tactics for managing, analyzing, mining, and visualizing data in order to understand the underlying patterns of variation. In this way, tens of thousands of individually curious participants are literally pushing new frontiers in science.

The Rewards of Feedback

Had a book like this been attempted before widespread use of the Internet, it would have been, at best, bland. It is no coincidence that the first regular appearances of the phrase "citizen science" in the 1990s coincided with the Internet's ascendance. During the previous era, when citizen participation in science was accomplished mainly via postal services, small bands of committed participants waited months or years before seeing any fruits of their labors—if they ever saw anything at all. Recruitment was difficult, and drop-out rates were high. Citizen science projects in those days rested on the backs of individuals already motivated by lifelong passions such as bird-watching or stargazing. For the most part, such individuals already knew enough about their field to trust that somebody, somewhere, might put their contributed observations to good use. Feedback was valued, but secondary to the observational process itself.

Every human habit, good or ill, feeds on reward. In the case of our great gift of curiosity, by definition the most sought-after rewards come in the form of discovery, new knowledge, and the "aha!" moments. Important for citizen science, discoveries need not be monumental in order to reward curiosity enough to elicit more of it. (As a preschooler sick at home, for example, simply discovering that a striking black-and-orange bird just outside my window exactly matched a picture of a male American Redstart in a book made me start searching for other such matches, thereby launching my lifelong passion for birds.) Almost as soon as the organizers of citizen science projects began adapting data-gathering protocols to the Internet, the best ones also recognized its spectacular capacity for acknowledging and thanking participants by providing background information related to the project, plus digests of patterns and analyses incorporating recent data. The "aha" moments were very powerful as participants began to see their own contributions fitting into the larger wholes of pattern and process across space and time. A near–real time, positively reinforcing loop was born—built on curiosity, fed by reward, and utterly impossible before the Internet. This loop has become the backbone of the most effective citizen science projects, because it fosters recruitment, retention, and viral expansion.

One Challenge Met, Others Beckon

If the operative noun in "citizen science" is about organized curiosity, then the primary challenge for the citizen science enterprise has been to produce genuine discoveries that could be described carefully, reviewed by peers, and published as advances in knowledge. Only through these processes can the outcome of any project qualify as science. Happily, a growing body of literature in technical scientific journals over the past 2 decades testifies that this challenge is being met (see Chapters 6–10). Indeed, there is little doubt that the breadth and depth of scientific discoveries will continue to increase as citizen science projects mature, observations accumulate, and analyses designed to mine large, messy data sets proliferate (Chapter 3).

With the "science" requirement increasingly secure, a few challenges pertaining to the "citizen" remain to be overcome. The most obvious is the need to strengthen the capacity of large Internet-based citizen science projects to support participants who wish to generate, and eventually answer, their own scientific questions. As currently configured, projects that engage participants exclusively as sensors to record and transmit observations were designed by scientists and organized to gather data that can be analyzed by specialists at the "back end." While many such projects accomplish genuine science, they do so by serving the curiosity of the project organizers rather than by fully exercising the curiosity of the participants themselves. As discussed above, the first step toward the latter involves providing accessible feedback to the participants in the form of creative graphics and interpretations that integrate the participants' data with key scientific questions driving the project. The next step is to provide direct and convenient access to the data themselves, so that participants and curious visitors can browse and answer their own questions directly from the project's results. The Map Room in the Great Backyard Bird Count (gbbc.birdsource.org/gbbcApps/maproom) represents an excellent example of this step. Even more challenging, and the ultimate step in fostering individual curiosity, will be to develop projects and tools through which participants can frame their own questions and design their own data-gathering protocols. As discussed by Trautman et al. (Chapter 12), this challenge currently represents a major goal of the advancing citizen science enterprise.

A far more daunting challenge facing purveyors of citizen science is to expand the participant base beyond the social pools already prone to exercising their curiosity. Everyone may be born to ask questions, but cultural norms and socioeconomic realities obviously impede exercising this gift for all but the most fortunate among us. In large part, the success of today's early-generation citizen science projects remains rooted in their appeal to avid hobbyists, not the masses. These projects may produce great data, and even startling discoveries, but the "dirty little secret" is that despite numerous promises to the contrary in public talks (my own included) and funded grant proposals (again, mine included), these projects have neither found the formula nor achieved the scale required to improve science literacy in any element of society. In theory, citizen science projects have enormous potential for providing culturally relevant gateways for urban and educationally deprived audiences around the world (Chapter 13), but none has yet succeeded. This failure should be viewed as unacceptable among any of us who strives to develop citizen science to its fullest potential. The social, economic, environmental, and even ethical consequences of abysmally sinking science literacy are so vast that any trick capable of reversing this trend needs to be explored fully. As an organized, collective expression for the ancient gift of natural human curiosity, citizen science may be the best such trick in the book. Certainly, the opportunity for testing this hypothesis is now upon us.

Contributing to Global Solutions in the Twenty-First Century

For the first time in the history of the earth, a system is becoming integrated by which one species (humans) can monitor the changing planet essentially in real time. Remote-sensing data from satellites contribute much of this information but cannot supply it all. Critical information about local conditions, plant phenology, water resources, distribution and behavior of organisms, and a host of other rapidly changing environmental variables can be supplied only by a broadly distributed network of human observers regularly communicating from local places. The Internet supplies the communication medium, and citizen science projects supply the tool kits for organizing and disseminating the information.

As the paces of landscape alterations and climate change accelerate through the twenty-first century, having sophisticated and multiscaled monitoring systems becomes increasingly urgent. Indeed, more and more scientists are recognizing the power of citizen science projects to gather time-sensitive information about environmental conditions, and even to document the effects of environmental calamities. A dramatic case in early twenty-first century U.S. history illustrates this point. The explosion and massive oil release at the Deepwater Horizon oil well in the Gulf of Mexico in the spring and summer of 2010 coincided with peaks of transgulf bird migration and nesting season for dozens of shorebird and seabird species along the beaches, marshes, and mangroves of the U.S. Gulf Coast. Within days of the disaster's onset, scientists, regulatory agencies, and conservationists called for information on the exact distribution and numbers of birds at risk, the ever-changing position of oil slicks, and the degree of oil contamination in both the habitats and the birds themselves. Almost overnight, two existing citizen science projects (eBird and NestWatch) created specialized tools and protocols by which local observers could input relevant observations. Federal agencies quickly made disaster-relief funds available to facilitate the process. During the early summer of 2010 alone, over 100,000 observations were submitted, documenting tens of thousands of nesting birds, including thousands with evidence of oil on their plumage but only a few hundred in which the oiling was extreme.

While the longer-term ecological consequences of the Deepwater Horizon disaster remain unclear at this writing, the immediate and large-scale monitoring of its specific, short-term impacts became a milestone achievement in citizen science. It is now clear that citizen science projects have come of age. They are informing the world about rapidly changing environmental conditions, and they are engaging increasingly large cohorts of citizens to learn more about our planet's ecosystems and how we might protect and restore them. It appears that humans have achieved the critical first steps toward a union between our individual curiosity and our collective power to document, interpret, and protect a complex and ever-changing world.

Literature Cited

Aiken, L.R. 2002. Attitudes and related psychosocial constructs: Theories, assessment, and research. Thousand Oaks, CA: Sage Publications.

Aiken, L.R., and D.R. Aiken. 1969. Recent research on attitudes concerning science. Science Education 53:295–305.

Aikenhead, G., and A. Ryan. 1992. The development of a new instrument: Views on science-technology-society (VOSTS). Science Education 76:477–491.

Ajzen, I. 1985. From intentions to actions: A theory of planned behavior. *In* J. Kuhl and J. Beckmann, eds., Action control: From cognition to behavior, pp. 11–39. Heidelberg: Springer-Verlag.

Altizer, S., W.M. Hochachka, and A.A. Dhondt. 2004. Seasonal dynamics of mycoplasmal conjunctivitis in eastern North American house finches. Journal of Animal Ecology 73:309–322.

Alwash, A. 2005. A meeting of the Marsh Arabs Forum. Nature Iraq 1 (1), http://www.natureiraq.org/site/sites/default/files/Nature%20vol%201–1.pdf (accessed 31 August 2009).

American Association for the Advancement of Science. 1993. Benchmarks for Science Literacy. New York: Oxford University Press.

Andrén, H. 1994. Effects of habitat fragmentation on birds and mammals in landscapes with different proportions of suitable habitat: A review. Oikos 71:355–366.

Andrewartha, H.G., and L.C. Birch. 1954. The distribution and abundance of animals. Chicago: University of Chicago Press.

Archer, M., M. Grantham, P. Howlett, and S. Stansfield. 2010. Bird observatories of the British Isles. London: T. & A.D. Poyser.

Armitage, D., R. Plummer, F. Berkes, R.I. Arthur, A.T. Charles, I.J. Davidson-Hunt, A.P. Diduck, N. Doubleday, D.S. Johnson, M. Marschke, et al. 2009. Adaptive management for social-ecological complexity. Frontiers in Ecology and the Environment 7:95–102.

Arnold, H.E., F.G. Cohen, and A. Warner. 2009. Youth and environmental action: Perspectives of young environmental leaders on their formative influences. Journal of Environmental Education 40:27–36.

Arnquist, S. 2009. Research trove: Patients' online data. New York Times, 25 August.

Arp, W., III, and C. Kenny. 1996. Black environmentalism in the local community context. Environment and Behavior 28:267–282.

Arvai, J.L., V.E.A. Campell, A. Baird, and L. Rivers. 2004. Teaching students to make better decisions about the environment: lessons from the decision sciences. Journal of Environmental Education 36:33–44.

Austin, G., and M.M. Rehfisch. 2005. Shifting non-breeding distributions of migratory fauna in relation to climatic change. Global Change Biology 11:31–38.

Backstrand, K. 2003. Civic science for sustainability: Reframing the role of experts, policy-makers and citizens in environmental governance. Global Environmental Politics 3:24–41.

Baillie, S.R. 1990. Integrated population monitoring of breeding birds in Britain and Ireland. Ibis 132:152–166.

Baillie, S.R., D.E. Balmer, I.S. Downie, and K.H.M. Wright. 2006. Migration watch: An Internet survey to monitor spring migration in Britain and Ireland. Journal of Ornithology 147:254–259.

Baillie, S.R., J.H. Marchant, D.I. Leech, A.C. Joys, D.G. Noble, C. Barimore, M.J. Grantham, K. Risely, and R.A. Robinson. 2009. Breeding birds in the wider countryside: Their conservation status 2008. BTO Research Report No. 516, http://www.bto.org/birdtrends. Thetford: British Trust for Ornithology.

Baillie, S.R., W.J. Sutherland, S.N. Freeman, R.D. Gregory, and E. Paradis. 2000. Consequences of large-scale processes for the conservation of bird populations. Journal of Applied Ecology 37:88–102.

Ballantyne, R., J. Fien, and J. Packer. 2000. Program effectiveness in facilitating intergenerational influence in environmental education: Lessons from the field. Journal of Environmental Education 32:8–15.

Ballantyne, R., J. Fien, and J. Packer. 2001. School environment education programme impacts upon student and family learning: A case study analysis. Environmental Education Research 7:23–37.

Basu, J.S., and A. Calabrese Barton. 2007. Developing a sustained interest in science among urban minority youth. Journal of Research in Science Teaching 44:466–489.

Batalden, R., K.S. Oberhauser, and A.T. Peterson. 2007. Ecological niches in breeding generations of Eastern North American monarch butterflies. Environmental Entomology 36:1365–1373.

Bauer, H.H. 1992. Scientific literacy and the myth of the scientific method. Urbana: University of Illinois Press.

Bauer, M., and I. Schoon. 1993. Mapping variety in public understanding of science. Public Understanding of Science 2:141–155.

Bauer, M.W., K. Petkova, and P. Boyadjieva. 2000. Public knowledge of and attitudes to science: Alternative measures that may end the "science war." Science, Technology, & Human Values 25:30–51.

Beck, U. 1991. Ecological enlightenment: Essays on the politics of the risk society. Atlantic Highlands, NJ: Humanities Press.

Beck, U. 1995. Ecological politics in an age of risk. Cambridge: Polity Press.

Becker, M.L., R.G. Congalton, R. Budd, and A. Fried. 1998. A GLOBE collaboration to develop land cover data collection and analysis protocols. Journal of Science Education and Technology 7:85–96.

Beeby, A. 1990. Toxic metal uptake and essential metal regulation in terrestrial invertebrates: A review. *In* M.C. Newman and A.W. McIntosh, eds., Metal ecotoxicology: Concepts and applications, p. 399. Chelsea, MI: Lewis Publishers.

Bell, J.F., J.F. Wilson, and G.C. Liu. 2008. Neighborhood greenness and 2-year changes in body mass index of children and youth. American Journal of Preventive Medicine 35:547–553.

Berkes, F., and C. Folke. 1998. Linking social and ecological systems. London: Cambridge University Press.

Berkowitz, A.R. 1996. A simple framework for considering the benefits of SSPs. National conference on student and scientist partnerships; 23–25 October 1996; Washington, DC: TERC and the Concord Consortium.

Berkowitz, A.R., M.E. Ford, and C.A. Brewer. 2005. A framework for integrating ecological literacy, civics literacy and environmental citizenship in environmental education. *In* E. Johnson and M. Mappin, eds., Environmental education and advocacy: Changing perspectives of ecology and education, pp. 227–266. Cambridge: Cambridge University Press.

Bhattacharjee, Y. 2005. Citizen scientists supplement work of Cornell researchers. Science 308: 1402–1403.

Bibby, C.J. 2003. Fifty years of Bird Study. Bird Study 50:194–210.

Biel, A. 2003. Environmental behavior: Changing habits in a social context. *In* A. Biel, B. Hannson, and M. Martensson, eds., Individual and structural determinants of environmental practice, pp. 11–25. London: Ashgate.

Birdlife International. 2009. Uncovering Iraq's unique wildlife, http://www.birdlife.org/news/news/2009/04/nature_iraq_surveys.html (accessed 31 August 2009). Cambridge, UK: Birdlife International.

Birnbaum, J.A.L. 2007. The development of natural resource values: Federal natural resource managers in Humboldt County. Master's thesis, Humboldt State University, Arcata, CA.

Bixler, R.D., C.L. Carlisle, W.E. Hammitt, and M.F. Floyd. 1994. Observed fears and discomforts among urban students on school field trips to wildland areas. Journal of Environmental Education 26:24–33.

Bixler, R.D., and M.F. Floyd. 1997. Nature is scary, disgusting, and uncomfortable. Environment and Behavior 29:443–467.

Bixler, R.D., M.F. Floyd, and W.E. Hammitt. 2002. Environmental socialization: Quantitative tests of the childhood play hypothesis. Environment and Behavior 34:795–818.

Blackburn, M., and K.J. Gaston, eds. 2003. Macroecology: Concepts and consequences. Oxford: Blackwell.

Blackmore, C. 2007. What kinds of knowledge, knowing and learning are required for addressing resource dilemmas? A theoretical overview. Environmental Science and Policy 10:512–525.

Bodzin, A.M. 2008. Integrating instructional technologies in a local watershed investigation with urban elementary learners. Journal of Environmental Education 39:47–57.

Bogner, F.X. 1998. The influence of short-term outdoor ecology education on long-term variable of environmental perspective. Journal of Environmental Education 29:17–29.

Bonney, R. 2004. Understanding the process of research. *In* D. Chittenden, G. Farmelo, and B.V. Lewenstein, eds., Creating connections: Museums and the public understanding of current research, pp. 199–210. Walnut Creek, CA: Alta-Mira Press.

Bonney, R. 2008. Citizen science at the Cornell Lab of Ornithology. *In* J. Falk, ed., Exemplary science in informal education settings, pp. 213–229. Arlington, VA: NSTA.

Bonney, R., H. Ballard, R. Jordan, E. McCallie, T. Phillips, J. Shirk, and C. Wilderman. 2009. Public participation in scientific research: Defining the field and assessing its potential for informal science education. A CAISE inquiry group report. Washington, DC: Center for Advancement of Informal Science Education (CAISE).

Bonney, R., C. Cooper, J. Dickinson, S. Kelling, T. Phillips, K. Rosenberg, and J. Shirk. 2009. Citizen science: A developing tool for expanding science knowledge and science literacy. BioScience 59:977–984.

Bonter, D., and W. Hochachka. 2003. Widespread declines of chickadees and corvids: Possible impacts of West Nile virus. American Birds 103rd Christmas Bird Count:22–25.

Bonter, D.N., and M.G. Harvey. 2008. Winter survey data reveal rangewide decline in Evening Grosbeak populations. Condor 110:376–381. doi:10.1525/cond.2008.8463.

Bonter, D.N., B. Zuckerberg, and J.L. Dickinson. 2010. Invasive birds in a novel landscape: Habitat associations and effects on established species. Ecography 33:494–502.

Borden, R.J. 1984/1985. Psychology and ecology: Beliefs in technology and the diffusion of ecological responsibility. Journal of Environmental Education 16:14–19.

Borer, E.T., E.W. Seabloom, M.B. Jones, and M. Schildhauer. 2009. Some simple guidelines for effective data management. Bulletin of the Ecological Society of America 90:205–214. doi:10.1890/0012–9623–90.2.205.

Borgman, C.L. 2007. Scholarship in the digital age: information, infrastructure, and the Internet. Cambridge, MA: MIT Press.

Bouillion, L.M., and L.M. Gomez. 2001. Connecting school and community with science learning: Real world problems and school-community partnerships as contextual scaffolds. Journal of Research in Science Teaching 38:878–889.

Boulinier, T., J.D. Nichols, J.E. Hines, J.R. Sauer, C.H. Flather, and K.H. Pollock. 1998. Higher temporal variability of forest breeding bird communities in fragmented landscapes. Proceedings of the National Academy of Sciences of the United States of America 95:7497–7501.

Boulinier, T., J.D. Nichols, J.E. Hines, J.R. Sauer, C.H. Flather, and K.H. Pollock. 2001. Forest fragmentation and bird community dynamics: Inference at regional scales. Ecology 82:1159–1169.

Boyd, D., and N. Elison. 2007. Social network sites: Definition, history, and scholarship. Journal of Computer-Mediated Communication 13:210–230.

Boykoff, J. 2007. From convergence to contention: United States mass media representations of anthropogenic climate change science. Royal Geographical Society 32:477–489.

Boykoff, M., and J. Boykoff. 2004. Balance as bias: Global warming and the US presige press. Global Environmental Change 14:125–136.

Bransford, J.D., A. Brown, and R. Cocking. 1999. How people learn: Brain, mind, experience, and school. Washington, DC: National Academy Press.

Breiman, L. 1996. Bagging predictors. Machine Learning 24:123–140.

Breiman, L. 2001. Random forests. Technical report, Statistics Department, University of California, Berkeley.

Brennan, J.M., D.J. Bender, T.A. Contreras, and L. Fahrig. 2002. Focal patch landscape studies for wildlife management: Optimizing sampling effort across scales. *In* J. Lui and W.W. Taylor, eds., Integrating landscape ecology into natural resource management, pp. 68–91. New York: Cambridge University Press.

Brew, A. 2003. Teaching and research: New relationships and their implications for inquiry-based teaching and learning in higher education. Higher Education Research and Development 22:3–18.

Brewer, C. 2002. Conservation education partnerships in schoolyard laboratories: A call back to action. Conservation Biology 16:577–579.

Brewer, C. 2006. Translating data into meaning: Education in conservation biology. Conservation Biology 20:689–691.

Brody, M., A. Bangert, and J. Dillon. 2008. Assessing learning in informal science contexts. National Research Council for Science Learning in Informal Environments Committee, http://www7.nationalacademies.org/bose/Brody_Commissioned_Paper.pdf.

Brossard, D., B. Lewenstein, and R. Bonney. 2005. Scientific knowledge and attitude change: The impact of a citizen science project. International Journal of Science Education 27:1099–1121.

Brotons, L., W. Thuiller, M.B. Araújo, and A.H. Hirzel. 2004. Presence-absence versus presence-only modelling methods for predicting bird habitat suitability. Ecography 27:437–448.

Brown, J.H., and B.A. Maurer. 1989. Macroecology: The division of food and space among species on continents. Science 243:1145–1150.

Brown, J.S., and A. Collins. 1989. Situated cognition and the culture of learning. Educational Researcher 18:32–42.

Brunt, J.W. 2000. Data management principles, implementation and administration. *In* W.K. Michener and J.W. Brunt, eds., Ecological data: Design, management, and processing, pp. 25–44. Cambridge: Blackwell Science.

Bures, S., and K. Weidinger. 2000. Estimation of calcium intake by Meadow Pipit nestlings in an acidified area. Journal of Avian Biology 31:426–429.

Bures, S., and K. Weidinger. 2003. Sources and timing of calcium intake during reproduction in flycatchers. Oecologia 137:634–641.

Burger, J., N. Messian, S. Patel, A. del Prado, and C. Anderson. 2004. What a coincidence! The effects of incidental similarity on compliance. Personality and Social Psychology Bulletin 30:35–43.

Burt, R.S. 1987. Social contagion and innovation: Cohesion versus structural equivalence. American Journal of Sociology 92:1287–1335.

Bybee, R.W. 1995. Achieving scientific literacy. Science Teacher 62:28–33.

Calabrese Barton, A. 2007. Science learning in urban settings. *In* S.K. Abell and N.G. Lederman, eds., Handbook of research on science education, pp. 319–343. Mahwah, NJ: Lawrence Erlbaum Associates.

Calabrese Barton A., T.J. Hindin, I.R. Contento, M. Trudeau, K. Yang, S. Hagiwara, and P. Koch. 2001. Underprivileged urban mothers' perspectives on science. Journal of Research in Science Teaching 38:688–711.

Caldiero, C.T. 2007. Crisis storytelling: Fisher's narrative paradigm and news reporting. American Communication Journal 9(1):2.

Caldow, R.W.G., and P.A. Racey. 2000. Large-scale processes in ecology and hydrology. Journal of Applied Ecology 37:6–12.

Callow, M. 2004. Identifying promotional appeals for targeting potential volunteers: An exploratory study on volunteering motives among retirees. International Journal of Nonprofit and Voluntary Sector Marketing 9:261–274.

Carrier, S.J. 2009. Environmental education in the schoolyard: Learning styles and gender. Journal of Environmental Education 40:2–12.

Caruana, R., M. Elhawary, A. Munson, M. Riedewald, D. Sorokina, D. Fink, W. Hochachka, and S. Kelling. 2006. Mining citizen science data to predict prevalence of wild bird species. In Proceedings of the ACM SIGKDD International Conference on Knowledge Discovery and Data Mining, Chicago, IL, pp. 909–915. New York: Association for Computing Machinery.

Centola, D.M., and M. Macy. 2007. Complex contagions and the weakness of long ties. American Journal of Sociology 113:702–734.

Challis, B.H. 1993. Spacing effects on cued-memory tests depend on level of processing. Journal of Experimental Psychology: Learning, Memory, and Cognition 19:389–396.

Chamberlain, D.E., R.J. Fuller, R.G.H. Bunce, J.C. Duckworth, and M. Shrubb. 2000. Changes in the abundance of farmland birds in relation to the timing of agricultural intensification in England and Wales. Journal of Applied Ecology 37:771–788.

Chamberlain, D.E., D.E. Glue, and M.P. Toms. 2009. Sparrowhawk *Accipiter nisus* presence and winter bird abundance. Journal of Ornithology 150:247–254. doi:10.1007/s10336–008–0344–4.

Chamberlain, D.E., A.G. Gosler, and D.E. Glue. 2007. Effects of the winter beech-mast crop on bird occurrence in British gardens. Bird Study 54:120–126.

Chamberlain, D.E., S. Gough, H. Vaughan, J.A. Vickery, and G.F. Appleton. 2007. Determinants of bird species richness in public green spaces. Bird Study 54:87–97.

Chamberlain, D.E., and J.D. Wilson. 2000. The contribution of hedgerow structure to the value of organic farms to birds. In N. Aebischer, A. Evans, P. Grice, and J. Vickery, eds., Ecology and conservation of lowland farmland birds, pp. 57–88. Tring: British Ornithologists' Union.

Chamberlain, D.E., A.M. Wilson, S.J. Browne, and J.A. Vickery. 1999. Effects of habitat type and management on the abundance of skylarks in the breeding season. Journal of Applied Ecology 36:856–870.

Chamberlain, D.E., J.D. Wilson, and R.J. Fuller. 1999. A comparison of organic and conventional farm systems in southern Britain. Biological Conservation 88:307–320.

Chawla, L. 1998a. Significant life experiences revisited: A review of research on sources of environmental sensitivity. Journal of Environmental Education 29:11–21.

Chawla, L. 1998b. Research methods to investigate significant life experiences: Review and recommendations. Environmental Education Research 4:383–397.

Chawla, L. 1999. Life paths into effective environmental action. Journal of Environmental Education 31:15–26.

Chawla, L. 2001. Significant life experiences revisited once again: Response to vol. 5(4) "Five critical commentaries of significant life experience research in environmental education." Environmental Education Research 7:451–461.

Chipeniuk, R. 1995. Childhood foraging as a means of acquiring competent human cognition about biodiversity. Environment and Behavior 27:490–512.

Clary, E.G., and M. Snyder. 1999. The motivations to volunteer: Theoretical and practical considerations. Current Directions in Psychological Science 8:156–159.

Clary, E.G., M. Snyder, R.D. Ridge, J. Copeland, A.A. Stukas, J. Haugen, and P. Miene. 1998. Understanding and assessing the motivations of volunteers: A functional approach. Journal of Personality and Social Psychology 74:1516–1530.

Clayton, S.D. 1993. Environmental identity: A conceptual and an operational definition. In S.D. Clayton and S. Opotow, eds., Identity and the natural environment: The psychological significance of nature, pp. 45–66. Cambridge, MA: MIT Press.

Cohen, M., R. Artz, R. Draxler, P. Miller, L. Poissant, D. Niemi, D. Ratte, M. Deslauriers, R. Duval, R. Laurin, et al. 2004. Modeling the atmospheric transport and deposition of mercury to the Great Lakes. Environmental Research 95:247–265.

Cohn, J.P. 2008. Citizen science: Can volunteers do real research? BioScience 58:192–197.

Collins, H.M., and T.J. Pinch. 1993. The golem. Cambridge: Cambridge University Press.

Commission for Environmental Cooperation. 2008. North American Monarch Conservation Plan. Montreal: CEC Secretariat.

Congalton, R.G., and M.L. Becker. 1997. Validating student data for scientific use: An example from the GLOBE program. In K.C. Cohen, ed., Internet links for science education: Student-scientist partnerships, pp. 133–156. New York: Plenum Press.

Cooper, C., J. Dickinson, T. Phillips, and R. Bonney. 2007. Citizen science as a tool for conservation in residential ecosystems. Ecology and Society 12:11, http://www.ecologyandsociety.org/vol12/iss12/art11/.

Cooper, C., J. Dickinson, T. Phillips, and R. Bonney. 2008. Science explicitly for nonscientists. Ecology and Society 13:r1, http://www.ecologyandsociety.org/vol13/iss2/resp1/.

Cooper, C.B., W.M. Hochachka, G. Butcher, and A.A. Dhondt. 2005. Seasonal and latitudinal trends in clutch size: Thermal constraints during laying and incubation. Ecology 86:2018–2031.

Cooper, C.B., W.M. Hochachka, and A.A. Dhondt. 2005. Latitudinal trends in within-year reoccupation of nest boxes and their implications. Journal of Avian Biology 36:31–39.

Cooper, C.B., W.M. Hochachka, T.B. Phillips, and A.A. Dhondt. 2006. Geographical and seasonal gradients in hatching failure in Eastern Bluebirds Sialia sialis reinforce clutch size trends. Ibis 148:221–230.

Cooper, C.B., and J. Mills. 2005. New software for quantifying incubation behavior from time-series recordings. Journal of Field Ornithology 76:352–356.

Corburn, J. 2003. Bringing local knowledge into environmental decision making. Journal of Planning Education and Research 22:420–433.

Corcoran, P.B. 1999. Formative influence in the lives of environmental educators in the United States. Environmental Education Research 5:207–220.

Cordell, H. 2004. Outdoor recreation for the 21st century: A report to the nation; The national survey on recreation and the environment. State College, PA: Venture Publishing.

Crawford, B.A. 2000. Embracing the essence of inquiry: New roles for science teachers. Journal of Research in Science Teaching 37:916–937.

Crick, H.Q.P. 1992. A bird-habitat coding system for use in Britain and Ireland incorporating aspects of land-management and human activity. Bird Study 39:1–12.

Crick, H.Q.P., and T.H. Sparks. 1999. Climate change related to egg-laying trends. Nature 399:423–424.

Cristianini, N., and J. Shawe-Taylor. 2000. An introduction to support vector machines and other kernel-based learning methods. Cambridge: Cambridge University Press.

Cronin-Jones, L.L. 2000. The effectiveness of schoolyards as sites for elementary science instruction. School Science and Mathematics 100:203–211.

Danielsen, F., N. Burgess, and A. Balmford. 2005. Monitoring matters: Examining the potential of locally-based approaches. Biodiversity and Conservation 14:2507–2542.

Dawson, B. 2009. Citizen journalists and citizens scientific redefine disaster story, http://www.sej.org/publications/disasters/citizens-journalist-and-citizens-scientific-redefine-disaster-story (accessed 2 July 2011). Jenkintown, PA: Society of Environmental Journalists.

De'ath, G. 2007. Boosted trees for ecological modeling and prediction. Ecology 88:243–251.

De'ath, G., and K. Fabricius. 2000. Classification and regression trees: A powerful yet simple technique for ecological data analysis. Ecology 81:3178–3192.

DeBoer, G.E. 2000. Scientific literacy: Another look at its historical and contemporary meanings and its relationship to science education reform. Journal of Research in Science Teaching 37:582–601.

Dettman-Easler, D., and J.L. Pease. 1999. Evaluating the effectiveness of residential environmental education programs in fostering positive attitudes toward wildlife. Journal of Environmental Education 31:33–39.

Dhondt, A.A. 1997. Negative data have positive value. Birdscope, 11:10–11.

Dhondt, A.A. 2007. What drives differences between North American and Eurasian tit studies? In K.A. Otter, ed., Ecology and behavior of chickadees and titmice: An integrated approach, pp. 299–310. Oxford: Oxford University Press.

Dhondt, A.A., S. Altizer, E.G. Cooch, A.K. Davis, A. Dobson, M.J.L. Driscoll, B.K. Hartup, D.M. Hawley, W.M. Hochachka, P.R. Hosseini, et al. 2005. Dynamics of a novel pathogen in an avian host: Mycoplasmal conjunctivitis in House Finches. Acta Tropica 94:77–93.

Dhondt, A.A., and W.M. Hochachka. 2001. Variations in calcium use by birds during the breeding season. Condor 103:592–598.

Dhondt, A.A., T.L. Kast, and P.E. Allen. 2002. Geographical differences in seasonal clutch size variation in multi-brooded bird species. Ibis 144:646–651.

Dhondt, A.A., D.L. Tessaglia, and R.L. Slothower. 1998. Epidemic mycoplasmal conjunctivitis in House Finches from eastern North America. Journal of Wildlife Diseases 34:265–280.

Diamond, J. 1999. Practical evaluation guide: Tools for museums and other informal educational settings. Walnut Creek, CA: AltaMira Press.

Diani, M.M., and D. McAdam. 2003. Social movements and social networks: Relational approaches to collective action. New York: Oxford University Press.

Dickinson, J. 2009. The people paradox: Self-esteem striving, immortality ideologies, and human response to climate change. Ecology and Society 14:34, http://www.ecologyandsociety.org/vol14/iss1/art34/.

Dickinson, J.L., B.Z. Zuckerberg, and D.N. Bonter. 2010. Citizen science as an ecological research tool: Challenges and benefits. Annual Review of Ecology, Evolution, and Systematics 41:149–172.

Donald, P.F., and R.J. Fuller. 1998. Ornithological atlas data: A review of uses and limitations. Bird Study 45:129–145.

Donovan, T.M., and C.H. Flather. 2002. Relationships among North American songbird trends, habitat fragmentation, and landscape occupancy. Ecological Applications 12:364–374.

Doran, G.T., A.F. Miller, and J.A. Cunningham. 1981. How to avoid costly job mismatches. Management Review, 70:35–36.

Dowling, J.L., D.A. Luther, and P.P. Marra. 2012. Comparative effects of urban development and anthropogenic noise on bird songs. Behavioral Ecology: in revision.

Drayton, B., and J. Falk. 2006. Dimensions that shape teacher-scientist collaborations for teacher enhancement. Science Education 90:734–761.

Drent, P.J., and J.W. Woldendorp. 1989. Acid rain and eggshells. Nature 339:431. doi:10.1038/339431a0.

Dresner, M., and M. Gill. 1994. Environmental education at summer nature camp. Journal of Environmental Education 25:35–41.

Driscoll, C.T., G.B. Lawrence, A.J. Bulger, T.J. Butler, C.S. Cronan, C. Eagar, K.F. Lambert, G.E. Likens, J.L. Stoddard, and K.C. Weathers. 2001. Acidic deposition in the northeastern United States: Sources and inputs, ecosystem effects, and management strategies. Bioscience 51:180–198.

Dunlap, J. 1996. Birds in the bushes: A story about Margaret Morse Nice. Minneapolis: Carolrhoda Books.

Dunlap, R. 1992. Trends in public opinion toward environmental issues: 1965–1990. In R. Dunlap and A. Mertig, eds., Trends in public opinion toward environmental issues: 1965–1990, pp. 89–116. Philadelphia: Taylor and Francis.

Dunlap, R., K. Van Liere, A. Mertig, and R.E. Jones. 2000. Measuring endorsement of the new ecological paradigm: A revised NEP scale. Journal of Social Issues 56:425–442.

Dunlap, R.E., and K.D. Van Liere. 1978. The "new environmental paradigm": A proposed measuring instrument and preliminary results. Journal of Environmental Education 9:10–19.

Dunn, E.H. 1993. Bird mortality from striking residential windows in winter. Journal of Field Ornithology 64:302–309.

Dunn, E.H., and D. Tessaglia. 1994. Predation of birds at feeders in winter. Journal of Field Ornithology 65:8–16.

Dunn, E.H., J. Bart, B.T. Collins, G. Craig, B. Dale, C.M. Downes, C.M. Francis, S. Woodley, and P. Zorn. 2006. Monitoring bird populations in small geographic areas. Special publication of the Canadian Wildlife Service, Environment Canada.

Dunn, P.O., and D.W. Winkler. 1999. Climate change has affected the breeding date of Tree Swallows throughout North America. Proceedings of the Royal Society of London, series B 266:2487–2490.

Durant, J., G. Evans, and G. Thomas. 1989. The public understanding of science. Nature 340:11–14.

Edelson, D.C., A. Tarnoff, K. Schwille, M. Bruozas, and A. Switzer. 2006. Learning how to make systematic decisions. Science Teacher (April).

Ehrenfeld, J.G., P. Kourtev, and W. Huang. 2001. Changes in soil functions following invasions of exotic understory plants in deciduous forests. Ecological Applications 11:1287–1300.

Elith, J., and C. Graham. 2009. Do they? How do they? WHY do they differ? On finding reasons for differing performances of species distribution models. Ecography 32:66–77.

Elith, J., C. Graham, R. Anderson, M. Dudik, S. Ferrier, A. Guisan, R. Hijmans, F. Huettmann, J. Leathwick, A. Lehmann, et al. 2006. Novel methods improve prediction of species' distributions from occurrence data. Ecography 29:129–151.

Engel, S.R., and J.R. Voshell. 2002. Volunteer biological monitoring: Can it accurately assess the ecological condition of streams? American Entomologist 48:164–177.

Ernstson, H., S.E. vanderLeeuw, C.L. Redman, D.J. Meffert, G. Davis, C. Alfsen, and T. Elmqvist. 2010. Urban transitions: On urban resilience and human-dominated ecosystems. Ambio 31:531–545.

Etheredge, S., and A. Rudnitsky. 2003. Introducing students to scientific inquiry: How do we know what we know? New York: Allyn and Bacon.

Evans, C., E. Abrams, R. Reitsma, K. Roux, L. Salmonsen, and P.P. Marra. 2005. The Neighborhood Nestwatch program: Participant outcomes of a citizen-science ecological research project. Conservation Biology 19:589–594.

Evans, C.A., E.D. Abrams, B.N. Rock, and S.L. Spencer. 2001. Student/scientist partnerships: A teacher's guide to evaluating the critical components. American Biology Teacher 63:318–323.

Evans, G., and J. Durant. 1995. The relationship between knowledge and attitudes in the public understanding of science in Britain. Public Understanding of Science 4:57–74.

Evans, G.W., G. Brauchle, A. Haq, R. Stecker, K. Wong, and E. Shapiro. 2007. Young children's environmental attitudes and behaviors. Environment and Behavior 39:635–659.

Evans, K.L., D.I. Leech, H.Q.P. Crick, J.J.D. Greenwood, and K.J. Gaston. 2009. Longitudinal and seasonal patterns in clutch size of some single-brooded British birds. Bird Study 56:75–85.

Evers, D.C., L.J. Savoy, C.R. DeSorbo, D.E. Yates, W. Hanson, K.M. Taylor, L.S. Siegel, J.H. Cooley, M.S. Bank, A. Major, et al. 2008. Adverse effects from environmental mercury loads on breeding common loons. Ecotoxicology 17:69–81. doi:10.1007/s10646–007–0168–7.

Ewert, A., G. Place, and J. Sibthorp. 2005. Early-life outdoor experiences and an individual's environmental attitudes. Leisure Sciences 27:225–239.

Faber Taylor, A., and F.E. Kuo. 2009. Children with attention deficits concentrate better after walk in the park. Journal of Attentional Disorders 12:402–409.

Fahrig, L. 2003. Effects of habitat fragmentation on biodiversity. Annual Review of Ecology Evolution and Systematics 34:487–515.

Fahrig, L. 2005. When is a landscape perspective important? In J. Wiens and M. Moss, eds., Issues and perspectives in landscape ecology, pp. 3–10. Cambridge: Cambridge University Press.

Falk, J.H., and L. Dierking. 2002. Lessons without limit: How free-choice learning is transforming education. Walnut Creek, CA: AltaMira Press.

Falk, J.H., and M. Storksdieck. 2009. Science learning in a leisure setting. Journal of Research in Science Teaching 47:194–212.

Farmer, J.F., D. Knapp, and G.M. Benton. 2007. An elementary school environmental education field trip: Long-term effects on ecological and environmental knowledge and attitude development. Journal of Environmental Education 38:33–42.

Ferguson, N. 2006. The next war of the world. Foreign Affairs 85 (5).

Fernandez-Gimenez, M.E., H.L. Ballard, and V.E. Sturtevant. 2008. Adaptive management and social learning in collaborative and community-based monitoring: A study of five community-based forestry organizations in the western USA. Ecology and Society 13 (2):4.

Ferrari, T.M., K.S. Lekies, and N. Arnett. 2009. Opportunities matter: Exploring youth's perspectives on their long-term participation in an urban 4-H youth development program. Journal of Youth Development 4 (3).

Field, D., P. Voss, T. Kuczenski, R. Hammer, and V. Radeloff. 2003. Reaffirming social landscape analysis in landscape ecology: A conceptual framework. Society & Natural Resources 16:349–361.

Fielding, A.H., and J.F. Bell. 1997. A review of methods for the assessment of prediction errors in conservation presence/absence models. Environmental Conservation 24:38–49.

Fink, D., and W.M. Hochachka. 2009. Gaussian semiparametric analysis using hierarchical predictive models. *In* D.L. Thomson, E.G. Cooch, and M.J. Conroy, eds., Modeling demographic processes in marked populations, pp. 1011–1036. New York: Springer.

Fink, D., W.M. Hochachka, B. Zuckerberg, D.W. Winkler, B. Shaby, M.A. Munson, G. Hooker, M. Riedewald, D. Sheldon, and S. Kelling. 2010. Spatiotemporal exploratory models for broad-scale survey data. Ecological Applications 20 (8): 2131–2147.

Fischer, J., D. Stallknecht, M. Luttrell, A. Dhondt, and K. Converse. 1997. Mycoplasmal conjunctivitis in wild songbirds: The spread of a new contagious disease in a mobile host population. Emerging Infectious Diseases 3:69–72.

Fishbein, M., and I. Ajzen. 1975. Belief, attitude, intention, and behavior: An introduction to theory and research. Reading, MA: Addison-Wesley.

Fitzpatrick, J.W., and F.B. Gill. 2002. Birdsource: Using birds, citizen science, and the Internet as tools for global monitoring. *In* J.N. Levitt, ed., Conservation in the Internet age: Threats and opportunities, pp. 165–185. London: Island Press.

Flather, C.H., and J.R. Sauer. 1996. Using landscape ecology to test hypotheses about large-scale abundance patterns in migratory birds. Ecology 77:28–35.

Folke, C., S. Carpenter, T. Elmqvist, L. Gunderson, C.S. Holling, B. Walker, J. Bengtsson, F. Berkes, J. Colding, K. Danell, et al. 2002. Resilience and sustainable development: Building adaptive capacity in a world of transformations. Report for the Environmental Advisory Council, Stockholm.

Forman, R.T.T. 1995. Land mosaics: The ecology of landscapes and regions. Cambridge: Cambridge University Press.

Fougere, M. 1998. The educational benefits to middle school students participating in a student/scientist partnership. Journal of Science Education and Technology 7:25–30.

Fraser, J., J. Sickler, A. Taylor, and S.D. Clayton. 2009. Belonging at the zoo: Retired volunteers, conservation activism and collective identity. Ageing and Society 29:351–368.

Frechtling, J.A., and L.M. Sharp. 1997. User-friendly handbook for mixed method evaluations. Directorate for Education and Human Resources, Division of Research, Evaluation, and Communication. National Science Foundation, Arlington, VA.

Freeman, S.N., D.G. Noble, S.E. Newson, and S.R. Baillie. 2007. Modelling population changes using data from different surveys: The Common Birds Census and the Breeding Bird Survey. Bird Study 54:61–72.

Freeman, S.N., R.A. Robinson, J.A. Clark, B.M. Griffin, and S.Y. Adams. 2007. Changing demography and population decline in the Common Starling *Sturnus vulgaris*: A multisite approach to integrated population monitoring. Ibis 149:587–596.

Friedman, A. 2008. Framework for evaluating impacts of informal science education projects, http://insci.org/resources/Eval_Framework.pdf. Washington, DC: National Science Foundation.

Friedman, J.H. 2001. Greedy function approximations: A gradient boosting machine. Annals of Statistics 29:1189–1232.

Friedman, T.L. 2005. The world is flat: A brief history of the twenty-first century. New York: Farrar, Straus and Giroux.

Fuller, R.J. 1982. Bird habitats in Britain. Calton: T. & A.D. Poyser.

Fuller, R.J., R.D. Gregory, D.W. Gibbons, J.H. Marchant, J.D. Wilson, S.R. Baillie, and N. Carter. 1995. Population declines and range contractions among lowland farmland birds in Britain. Conservation Biology 9:1425–1441.

Fuller, R.J., D.G. Noble, K.W. Smith, and D. Vanhinsbergh. 2005. Recent declines in populations of woodland birds in Britain: A review of possible causes. British Birds 98:116–143.

Fuller, R.J., L.R. Norton, R.E. Feber, P.J. Johnson, D.E. Chamberlain, A.J. Joys, F. Mathews, R.C. Stuart, M.C. Townsend, W.J. Manley, et al. 2005. Benefits of organic farming to biodiversity vary among taxa. Biology Letters 1:431–434.

Garibay, C. 2004. Museums and community outreach: A secondary analysis of a museum collaborative program. San Francisco: Saybrook Graduate School and Research Center.

Garibay, C. 2009. Urban bird gardens: Front-end research. Cornell Lab of Ornithology, Ithaca, NY.

Gaston, K.J. 2003. The structure and dynamics of geographic ranges. Oxford: Oxford University Press. (Oxford series in ecology and evolution).

Geller, E.S. 1995. Integrating behaviorism and humanism for environmental protection. Journal of Social Issues 51:179–195.

Genet, K.S., and L.G. Sargent. 2003. Evaluation of methods and data quality from a volunteer-based amphibian call survey. Wildlife Society Bulletin 31:703–714.

Gerlach, C., I. Law, A. Gade, and O.B. Paulson. 1999. Perceptual differentiation and category effects in normal object recognition: A PET study. Brain 122:2159–2170.

Gibbons, D.W., P.F. Donald, H.-G. Bauer, L. Fornasari, and I.K. Dawson. 2007. Mapping avian distributions: The evolution of bird atlases. Bird Study 54:324–334.

Gibbons, D.W., J.B. Reid, and R.A. Chapman. 1993. The new atlas of breeding birds in Britain and Ireland: 1988–1991. London: T. & A.D. Poyser.

Gick, M.L., and K.J. Holyoak. 1980. Analogical problem solving. Cognitive Psychology 12:306–355.

Gillings, S., and R.J. Fuller. 2001. Habitat selection by Skylarks *Alauda arvensis* wintering in Britain in 1997/98. Bird Study 48:293–307. doi:10.10 80/00063650109461229.

Godin, B., and Y. Gingras. 2000. What is scientific and technological culture and how is it measured? A multidimensional model. Public Understanding of Science 9:43–58.

Goldstone, R.L. 1998. Perceptual learning. Annual Review of Psychology 49:585–612.

González, N., and L.C. Moll. 2002. Cruzando el puente: Building bridges to funds of knowledge. Educational Policy 16:623–641.

Gottschalk, T.K., F. Huettmann, and M. Ehlers. 2005. Thirty years of analysing and modelling avian habitat relationships using satellite imagery data: A review. International Journal of Remote Sensing 26:2631–2656.

Gough, A. 1999. Kids don't like wearing the same jeans as their mums and dads: So whose "life" should be in significant life experiences research? Environmental Education Research 5:383–394.

Gough, N. 1999. Surpassing our own histories: Autobiographical methods for environmental education research. Environmental Education Research 5:407–418.

Gough, S. 1999. Significant life experiences (SLE) research: A view from somewhere. Environmental Education Research 5:353–363.

Grace, J.B. 2008. Structural equation modeling for observational studies. Journal of Wildlife Management 72:14–22. doi:10.2193/2007-307.

Granovetter, M. 1983. The strength of weak ties: A network theory revisited. Sociological Theory 1:201–233.

Granovetter, M. 1985. Economic action and social structure: The problem of embeddedness. American Journal of Sociology 91:481–510.

Graveland, J. 1995. Arthropods and seeds are not sufficient as calcium sources for shell formation and skeletal growth. In The quest for calcium: Calcium limitation in the reproduction of forest paserines in relation to snail abundance and soil acidification. PhD thesis, Rijksuniversiteit Groningen, Netherlands, p. 169.

Graveland, J. 1996. Calcium deficiency in wild birds. Veterinary Quarterly 18:S136–S137.

Graveland, J. 1998. Effects of acid rain on bird populations. Environmental Reviews 6:41–54.

Graveland, J. and R.H. Drent, 1997. Calcium availability limits breeding success of passerines on poor soils. Journal of Animal Ecology 66:279–288.

Graveland, J., R. Van der Wal, J.H. Van Balen, and A.J. Van Noordwijk. 1994. Poor reproduction in forest passerines from decline of snail abundance on acidified soils. Nature 368:446–448.

Greene, J.C., V.J. Caracelli, and W.F. Graham. 1989. Toward a conceptual framework for mixed-method evaluation design. Educational Evaluation and Policy Analysis 11:255–274.

Greenwood, J.J.D. 2007a. Citizens, science and bird conservation. Journal of Ornithology 148:S77–S124.

Greenwood, J.J.D. 2007b. Nicholson, (Edward) Max (1904–2003). Oxford Dictionary of National Biography, http://www.oxforddnb.com/. Oxford: Oxford University Press.

Greenwood, J. 2009. 100 years of ringing in Britain and Ireland. Ringing & Migration 24:147–153.

Greenwood, J.J.D., and R.A. Robinson. 2007. Principles of sampling. In W.J. Sutherland, ed. Ecological censusing techniques: A handbook, 2nd ed., pp. 11–86. Cambridge: Cambridge University Press.

Gregory, J., and S. Miller. 1998. Science in public: Communication, culture, and credibility. Cambridge, MA: Perseus.

Griffing, S.M., A.M. Kilpatrick, L. Clark, and P.P. Marra. 2007. Mosquito landing rates on nesting American robins (Turdus migratorius). Vector-Borne and Zoonotic Diseases 7:437–443. doi:10.1089/vbz.2006.0560.

Guegan, N., and A. Martin. 2009. Incidental similarity facilitates behavioral mimicry. Social Psychology 40:88–92.

Hagemeijer, W.J.M., and M.J. Blair, eds. 1997. The EBCC atlas of European breeding birds: Their distribution and abundance. London: T & A.D. Poyser.

Haladyna, T., and J. Shaughnessy. 1982. Attitudes toward science: A quantitative synthesis. Science Education 66:547–563.

Hames, R.S. 2001. Habitat fragmentation and forest birds: Effects at multiple scales. PhD dissertation, Cornell University, Ithaca, NY, p. 198.

Hames, R.S., J.D. Lowe, and C.F. David. 2010. A categorical method of estimating biomass of calcium-rich invertebrates available to breeding birds in eastern hardwood forests. Cornell Lab of Ornithology, Ithaca, NY.

Hames, R.S., J.D. Lowe, S. Barker Swarthout, and K.V. Rosenberg. 2006. Understanding the risk to neotropical migrant bird species of multiple human-caused stressors: Processes behind the patterns. Ecology and Society 11:24.

Hames, R.S., K.V. Rosenberg, J.D. Lowe, S.E. Barker, and A.A. Dhondt. 2002a. Adverse effects of acid rain on the distribution of the Wood Thrush *Hylocichla mustelina* in North America. Proceedings of the National Academy of Sciences of the United States of America 99:11235–11240.

Hames, R.S., K.V. Rosenberg, J.D. Lowe, S.E. Barker, and A.A. Dhondt. 2002b. Effects of forest fragmentation on North American tanager and thrush species in eastern and western landscapes. Studies in Avian Biology 25:81–92.

Hansen, L.A. 1998. Where we play and who we are. Illinois Parks and Recreation (1998): 22–25.

Hanski, I.A., and M.E. Gilpin. 1997. Metapopulation biology: Ecology, genetics, and evolution. San Diego: Academic Pres.

Hardin, G. 1968. The tragedy of the commons. Science 162:1243–1248.

Harnik, P.G., and R.M. Ross. 2003a. Assessing data accuracy when involving students in authentic paleontological research. Journal of Geoscience Education 51:76–84.

Harnik, P.G., and R.M. Ross. 2003b. Developing effective K–16 geoscience research partnerships. Journal of Geoscience Education 51:5–8.

Harrell, F.E., Jr. 2001. Regression modeling strategies with applications to linear models, logistic regression and survival analysis. New York: Springer-Verlag.

Harris Poll. 2009. Half of Americans don't use Twitter, MySpace, Facebook (April). New York, NY: Harris Interactive.

Harris, L.D. 1984. The fragmented forest: Island biogeography theory and the preservation of biotic diversity. Chicago: University of Chicago Press.

Hartel, D.R. 2006. Gulf Coast tree assessment. Presentation to the International Society of Arboriculture at their 82nd Annual Conference & Trade Show, 29 July–2 August 2006; Minneapolis.

Hartup, B., A. Dhondt, K. Sydenstricker, W. Hochachka, and G. Kollias. 2001. Host range and dynamics of mycoplasmal conjunctivitis among birds in North America. Journal of Wildlife Diseases 37:72–81.

Hartup, B.K., H.O. Mohammed, G.V. Kollias, and A.A. Dhondt. 1998. Risk factors associated with mycoplasmal conjunctivitis in House Finches. Journal of Wildlife Diseases 34:281–288.

Haselton, M.G., and D. Nettle. 2006. The paranoid optimist: An integrative evolutionary model of cognitive biases. Personality and Social Psychology Review 10:47–66.

Hassel, C. 2004. Can diversity extend to ways of knowing? Engaging cross-cultural paradigms. Journal of Extension 42 (2), http://www.joe.org/joe/2004april/a7.shtml (accessed 31 March 2008).

Hastie, T., R. Tibshirani, and J.H. Friedman. 2001. The elements of statistical learning: Data mining, inference, and prediction. New York: Springer-Verlag.

Hauert, C., S. De Monte, J. Hofbauer, and K. Sigmund. 2002. Volunteering as a red queen mechanism for cooperation in public goods games. Science 296:1129–1132.

Hawkins, J.W., M.W. Lankester, and R.R.A. Nelson. 1998. Sampling terrestrial gastropods using cardboard sheets. Malacologia 39:1–9.

Heimlich, R.E., and W.D. Anderson. 2001. Development at the urban fringe and beyond: Impacts on agriculture and rural land. Economic Research Service, U.S. Department of Agriculture, Agricultural economic report no. 803, Washington, DC.

Henderson, I.G., J. Cooper, R.J. Fuller, and J. Vickery. 2000. The relative abundance of birds on set-aside and neighbouring fields in summer. Journal of Applied Ecology 37:335–347.

Hewson, C.M., A. Amar, J.A. Lindsell, R.M. Thewlis, S. Butler, K.W. Smith, and R.J. Fuller. 2007. Recent changes in bird populations in British broadleaved woodland. Ibis 149:14–28.

Hewson, C.M., and D.G. Noble. 2009. Population trends of breeding birds in British woodlands over a 32-year period: Relationships with food, habitat use and migratory behaviour. Ibis 151:464–486. doi:10.1111/j.1474–919X.2009.00937.x.

Hey, T., S. Tansley, and K. Tolle, eds. 2009. The fourth paradigm: Data-intensive scientific discovery. Redmond, WA: Microsoft Research.

Hickling, R., ed. 1983. Enjoying ornithology. Calton: T. & A.D. Poyser.

Hickling, R., D.B. Roy, J.K. Hill, and C.D. Thomas. 2005. A northward shift of range margins in British Odonata. Global Change Biology 11:502–506.

Hiemstra, C., G.E. Liston, R.A. Pielke, Sr., D.L. Birkenheuer, and S.C. Albers. 2006. Comparing local analysis and prediction system (LAPSO) assimilations with independent observations. Weather and Forecasting 21:1024–1042.

Higgins, R. 1993. Race and environmental equity: An overview of the environmental justice issue in the policy process. Polity 26:281–300.

Hines, J. 2008. PRESENCE. Version 2.0. Computer software program. Patuxent, MD: USGS.

Hitch, A.T., and P.L. Leberg. 2007. Breeding distributions of North American bird species moving north as a result of climate change. Conservation Biology 21:534–539. doi:10.1111/j.1523–1739.2006.00609.x.

Hmelo-Silver, C.E., and M.G. Pfeffer. 2004. Comparing expert and novice understanding of a complex system from the perspective of structures, behaviors, and functions. Cognitive Science 28:127–138.

Hochachka, W., J. Wells, K. Rosenberg, D. Tessaglia-Hymes, and A. Dhondt. 1999. Irruptive migration of Common Redpolls. Condor 101:195–204.

Hochachka, W.M., R. Caruana, D. Fink, A. Munson, M. Riedewald, D. Sorokina, and S. Kelling. 2007. Data-mining discovery of pattern and process in ecological systems. Journal of Wildlife Management 71:2427–2437.

Hochachka, W.M., and A.A. Dhondt. 2000. Density-dependent decline of host abundance resulting from a new infectious disease. Proceedings of the National Academy of Sciences of the United States of America 97:5303–5306.

Hoffman, S.M., and A. Oliver-Smith, eds. 2002. Catstrophe and culture: The anthropology of disaster. Oxford: James Curry.

Hooker, G. 2007. Generalized functional ANOVA diagnostics for high-dimensional functions of dependent variables. Journal of Computational and Graphical Statistics 16:709–732. doi:10.1198/106186007x237892.

Hosmer, D.W., and S. Lemeshow. 2000. Applied logistic regression, 2nd ed. New York: John Wiley & Sons.

Howe, J. 2008. Crowd sourcing: Why the power of the crowd is driving the future of business. New York: Crown.

Hurd, P.D. 1998. Scientific literacy: New minds for a changing world. Science Education 82:407–416.

IFRC. 2009a. World Disasters Report. Geneva: The International Federation of the Red Cross and Red Crescent Societies.

IFRC. 2009b. Disaster management: About disasters. Geneva: The International Federation of the Red Cross and Red Crescent Societies.

Irwin, A. 1995. Citizen science: A study of people, expertise and sustainable development. London: Routledge.

Jackson, S.F., G.E. Austin, and M.J.S. Armitage. 2006. Surveying waterbirds away from major waterbodies: Implications for waterbird population estimates in Great Britain. Bird Study 53:105–111.

Jacobson, S.K., M.D. McDuff, and M.C. Monroe. 2007. Promoting conservation through the arts: Outreach for hearts and minds. Conservation Biology 21:7–10.

James, J.J., and R.D. Bixler. 2008. Children's role in meaning making through their participation in an environmental education program. Journal of Environmental Education 39:44–59.

Jeanpierre, B., K. Oberhauser, and C. Freeman. 2005. Characteristics of professional development that effect change in secondary science teachers' classroom practices. Journal of Research in Science Teaching 42:668–690.

Jiguet, F., R. Julliard, C.D. Thomas, O. Dehorter, S.E. Newson, and D. Couvet. 2006. Thermal range predicts bird population resilience to extreme high temperatures. Ecology Letters 9:1321–1330.

Johnson, C., J. Bowker, and H.K. Cordell. 2004. Ethnic variation in environmental belief and behavior: An examination of the new ecological paradigm in a social psychological context. Environment and Behavior 3:157–186.

Johnson, L.S., and R.M.R. Barclay. 1996. Effects of supplemental calcium on the reproductive output of a small passerine bird, the House Wren (Troglodytes aedon). Canadian Journal of Zoology-Revue Canadienne de Zoologie 74:278–282.

Johnston, R.F. 1954. Variation in breeding season and clutch size in Song Sparrows on the Pacific coast. Condor 56:268–273.

Jones, M.B., M.P. Schildhauer, O.J. Reichman, and S. Bowers. 2006. The new bioinformatics: Integrating ecological data from the gene to the biosphere. Annual Review of Ecology, Evolution, and Systematics 37:519–544. doi:10.1146/annurev.ecolsys.37.091305.110031.

Jordan, R.C., D. Howe, W.R. Brooks, and J.G. Ehrenfeld. 2011. Volunteer accuracy in a citizen science project. Rutgers University, New Brunswick, NJ.

Józsa, G.I.G., M.A. Garrett, T.A. Oosterloo, H. Rampadarath, Z. Paragi, H. van Arkel, C. Lintott, W.C. Keel, K. Schawinski, and E. Edmondon. 2009. Revealing Hanny's Voorwerp: Radio observations of IC 2497. Astronomy & Astrophysics, 500 (2): L33–L36.

Kahle, J., and J. Meece. 1994. Research on girls in science lessons and applications. In D. Gabel, ed., Handbook of research in science teaching and learning, pp. 1559–1610. Washington, DC: National Science Teachers Association.

Kahneman, D., A.B. Krueger, D. Schkade, N. Schwartz, and A.A. Stone. 2006. Would you be happier if you were richer? A focusing illusion. Science 312:1908–1910.

Kallerud, E., and I. Ramberg. 2002. The order of discourse in surveys of public understanding of science. Public Understanding of Science 11:213–224.

Kals, E., D. Schumacher, and L. Montada. 1999. Emotional affinity toward nature as a motivational basis to protect nature. Environment and Behavior 31:178–202.

Kellert, S.R. 2002. Experiencing nature: Affective, cognitive, and evaluative development in children. *In* P.H. Kahn and S.R. Kellert, eds., Children and nature: psychological, sociocultural and evolutionary investigations, pp. 117–151. Cambridge, MA: MIT Press.

Kelling, S. 2008. Significance of organism observations: Data discovery and access in biodiversity research. Report for the Global Biodiversity Information Facility, Copenhagen, http://www.gbif.org/orc/?doc_id=1642.

Kelling, S., W.M. Hochachka, D. Fink, M. Riedewald, R. Caruana, G. Ballard, and G. Hooker. 2009. Data-intensive science: A new paradigm for biodiversity studies. BioScience 59:613–620. doi:10.1525/bio.2009.59.7.12.

Kendall, W.L., B.G. Peterjohn, and J.R. Sauer. 1996. First-time observer effects in the North American Breeding Bird Survey. Auk 113:823–829.

Kéry, M., J.A. Royle, and H. Schmid. 2005. Modeling avian abundance from replicated counts using binomial mixture models. Ecological Applications 15:1450–1461. doi:10.1890/04–1120.

Kéry, M., J.A. Royle, and H. Schmid. 2008. Importance of sampling design and analysis in animal population studies: A comment on Sergio et al. Journal of Applied Ecology 45:981–986. doi:10.1111/j.1365–2664.2007.01421.x.

Kéry, M., and H. Schmid. 2004. Monitoring programs need to take into account imperfect species detectability. Basic and Applied Ecology 5:65–73.

Klandermans, B. 2002. How group identification helps to overcome the dilemma of collective action. American Behavioral Scientist 45:887–900.

Kleinberg, J. 2000. Navigation in a small world. Nature 406:845.

Kollmuss, A., and J. Agyeman. 2002. Mind the gap: Why do people act environmentally and what are the barriers to pro-environmental behavior? Environmental Education Research 8:239–260.

Koriat, A., M. Goldsmith, and A. Pansky. 2000. Toward a psychology of memory accuracy. Annual Review of Psychology 51:481–537.

Kountoupes, D., and K.S. Oberhauser. 2008. Citizen science and youth audiences: Educational outcomes of the Monarch Larva Monitoring Project. Journal of Community Engagement and Scholarship 1:10–20.

Kovan, J., and J. Dirkx. 2003. "Being called awake": The role of transformative learning in the lives of environmental activists. Adult Education Quarterly 53:99–118.

Krackhardt, D. 1992. The strength of strong ties: The importance of philos in organizations. *In* N. Nohria and R.G. Eccles, eds., Networks and organizations: Structure, form, and action, pp. 216–239. Boston: Harvard Business School Press.

Krajcik, J., K.L. McNeill, and B.J. Reiser. 2008. Learning-goals-driven design model: Developing curriculum materials that align with national standards and incorporate project-based pedagogy. Science Education 92:1–32.

Krasny, M.E., and R. Bonney. 2005. Environmental education through citizen science and participatory action research. *In* E. Johnson and M. Mappin, eds., Environmental education and advocacy, pp. 292–319. Cambridge: Cambridge University Press.

Krasny, M.E., and K.G. Tidball. 2009. Applying a resilience systems framework to urban environmental education. Environmental Education Research 15:465–482.

Krebs, C.J. 2009. Ecology: The experimental analysis of distribution and abundance, 6th ed. San Francisco: Benjamin Cummings.

Kuo, F.E., and A. Faber Taylor. 2004. A potential natural treatment for attention-deficit/hyperactivity disorder: Evidence from a national study. American Journal of Public Health 94:1580–1586.

Lack, D. 1943. The life of the Robin. London: Witherby.

Lack, D. 1947. The significance of clutch-size. Ibis 89:302–352.

Lack, P. 1986. The atlas of wintering birds in Britain and Ireland. Calton: T. & A.D. Poyser.

Lack, P. 1992. Birds on lowland farms. London: HMSO.

LaDeau, S.L., A.M. Kilpatrick, and P.P. Marra. 2007. West Nile virus emergence and large-scale declines of North American bird populations. Nature 447:710–713. doi:10.1038/nature05829.

Lakshminarayanan, S. 2007. Using citizens to do science versus citizens as scientists. Ecology and Society 12:r2.

Lamb, D., and V. Bowersox. 2000. The national atmospheric deposition program: An overview. Atmospheric Environment 34:1661–1663.

Laugksch, R.C. 2000. Scientific literacy: A conceptual overview. Science Education 84:71–94.

Lave, J., and E. Wenger. 1991. Situated learning: Legitimate peripheral participation. Cambridge: University of Cambridge Press.

Lawless, J.G., and B.N. Rock. 1998. Student scientist partnerships and data quality. Journal of Science Education and Technology 7:5–13.

Lawrence, A. 2006. "No personal motive?" Volunteers, biodiversity, and the false dichotomies of participation. Ethics, Place & Environment 9:279–298.

Lawrenz, F. 1976. The prediction of student attitude toward science from student perception of the classroom learning environment. Journal of Research in Science Teaching 13:509–515.

Lederman, N.G. 1998. The state of science education: Subject matter without context. Electronic Journal of Science Education 3 (2), http://wolfweb.unr.edu/homepage/jcannon/ejse/ejsev3n2.html (accessed 4 July 2011).

Lederman, N.G., F. Abd-El-Khalick, R.L. Bell, and R.S. Schwartz. 2002. Views of nature of science questionnaire: Toward valid and meaningful assessment of learners' conceptions of nature of science. Journal of Research in Science Teaching 39:497–522.

Lederman, N.G., R.S. Schwartz, F. Abd-El-Khalick, and R.L. Bell. 2001. Pre-service teachers' understanding and teaching of the nature of science: An intervention study. Canadian Journal of Science, Mathematics, and Technology Education 1:135–160.

Lee, J.H., H. Hassan, G.E. Hill, E.W. Cupp, T.B. Higazi, C.J. Mitchell, M.S. Godsey, and T.R. Unnasch. 2002. Identification of mosquito avian-derived blood meals by polymerase chain reaction-heteroduplex analysis. American Journal of Tropical Medicine and Hygiene 66:599–604.

Leeming, F.C., W.O. Dwyer, and B.A. Bracken. 1995. Children's environmental attitude and knowledge scale: Construction and validation. Journal of Environmental Education 26:22–31.

Leeming, F.C., B.E. Porter, W.O. Dwyer, M.K. Cobern, and D.P. Oliver, 1997. Effects of participation in class activities on children's environmental attitudes and knowledge. Journal of Environmental Education 28:33–42.

Leitão, A.B. 2006. Measuring landscapes: A planner's handbook. Washington, DC: Island Press.

Lemoine, N., H.G. Bauer, M. Peintinger, and K. Böhning-Gaese. 2007. Effects of climate and land-use change on species abundance in a central European bird community. Conservation Biology 21:495–503.

Lepage, D., S. Kelling, and G. Ballard. 2005. The bird monitoring data exchange schema, http://www.avianknowledge.net/content/about/bird-monitoring-data-exchange (accessed 4 July 2011). Ithaca, NY: Cornell Lab of Ornithology.

Lerman, S.B., and P.S. Warren. 2011. The conservation value of residential landscapes: Exploring the links between birds and people. Ecological Applications 21:1327–1339.

Lindenmayer, D., and J. Fischer. 2006. Habitat fragmentation and landscape change: An ecological and conservation synthesis. Collingwood, Victoria: CSIRO.

Lindenmayer, D.B., and J. Fischer. 2007. Tackling the habitat fragmentation panchreston. Trends in Ecology & Evolution 22:127–132. doi:10.1016/j.tree.2006.11.006.

Lindsey, E., M. Mudresh, V. Dhulipala, K. Oberhauser, and S. Altizer. 2009. Crowding and disease: Effects of host density on response to infection in a butterfly-parasite interaction. Ecological Entomology 34:551–561.

Link, W.A., and J.R. Sauer. 2002. A hierarchical analysis of population change with application to Cerulean Warblers. Ecology 83:2832–2840.

Liu, J., and W.W. Taylor. 2002. Integrating landscape ecology into natural resource management. Cambridge: Cambridge University Press.

Lohr, V.I., and C.H. Pearson-Mims. 2005. Children's active and passive interactions with plants influence their attitudes and actions toward trees and gardening as adults. HortTechnology 15:472–476.

Losey, J.E., J.E. Perlman, and E.R. Hoebeke. 2007. Citizen scientists rediscover rare nine-spotted lady beetle, *Coccinella novemnotata,* in eastern North America. Journal of Insect Conservation 11:415–417.

Louv, R. 2005. Last child in the woods: Saving our children from nature-deficit disorder. Chapel Hill: Algonquin Books.

Ludescher, B., I. Altintas, S. Bowers, J. Cummings, T. Critchlow, E. Deelman, D. De Roure, J. Freier, C. Goble, M.B. Jones, et al. 2009. Scientific process automation and workflow management. *In* A. Shoshani and D. Rotem, eds., Scientific data mangement challenges, technology, and deployment, pp. 467–508. Boca Raton: CRC Press.

MacArthur, R.H., and E.O. Wilson. 1967. The theory of island biogeography. Princeton, NJ: Princeton University Press.

Machlis, G., J. Force, and W.J. Burch. 1997. The human ecosystem part 1: The human ecosystem as an organizing concept in ecosystem management. Society & Natural Resources 10:347–367.

MacKenzie, D.I. 2006. Occupancy estimation and modeling: Inferring patterns and dynamics of species. Burlington, MA: Elsevier.

MacKenzie, D.I., J.D. Nichols, J.A. Royle, K.H. Pollock, L.L. Bailey, and J.E. Hines. 2006. Occupancy estimation and modeling. Burlington, MA: Academic Press.

MacKenzie, D.I., J.D. Nichols, N. Sutton, K. Kawanishi, and L.L. Bailey. 2005. Improving inferences in popoulation studies of rare species that are detected imperfectly. Ecology 86:1101–1113.

Mackenzie, D.I., and J.A. Royle. 2005. Designing occupancy studies: General advice and allocating survey effort. Journal of Applied Ecology 42:1105–1114.

Macleod, R., P. Barnett, J. Clark, and W. Cresswell. 2005. Body mass change strategies in blackbirds *Turdus merula:* The starvation–predation risk trade-off. Journal of Animal Ecology 74:292–302.

Marchant, J.H., R. Hudson, S.P. Carter, and P. Whittington. 1990. Population trends in British breeding birds. Tring: British Trust for Ornithology.

Marcinkowski, T. 1993. Assessment in environmental education. *In* R.J. Wilke, ed., Environmental education teacher resource handbook, pp. 143–197. Thousand Oaks, CA: Corwin Press.

Marris, E. 2010. Birds flock online. Nature. doi:10.1038/news.2010.395.

Martin, E., B.G. Peterjohn, and S. Kelling. 2009. Gathering, organizing, and accessing data for use in bird conservation across the Americas. *In* Proceedings of the Fourth International Partners in Flight Conference: Tundra to Tropics, pp. 388–396. McAllen, TX: Partners in Flight.

Marzluff, J.M., R. Bowman, and R.E. Donnelly. 2001. A historical perspective on urban bird research: Trends, terms, and approaches. *In* J.M. Marzluff, R. Bowman, and R.E. Donnelly, eds., Avian conservation and ecology in an urbanizing world, pp. 1–17. Norwell, MA: Klewur Academic.

Mason, W., A. Jones, and R. Goldstone. 2008. Propagation of innovations in networked groups. Journal of Experimental Psychology 137:422–433.

Mathew, S., and R. Boyd. 2009. When does optional participation allow the evolution of cooperation? Proceedings of the Royal Society Series B 276:1167–1174.

McCormick, S., P. Brown, and S. Zavestoski. 2003. The personal is scientific, the scientific is political: The public paradigm of the environmental breast cancer movement. Sociological Forum 18:545–576.

McCoy, K.D. 1999. Sampling terrestrial gastropod communities: Using estimates of species richness and diversity to compare two methods. Malacologia 41:271–281.

McDavid, J.C., and L.R.L. Hawthorn. 2006. Program evaluation and performance measurement: An introduction to practice. Thousand Oaks, CA: Sage Publications.

McGarigal, K., and S.A. Cushman. 2002. Comparative evaluation of experimental approaches to the study of habitat fragmentation effects. Ecological Applications 12:335–345.

McGarigal, K., and S.A. Cushman. 2005. The gradient concept of landscape structure. *In* J. Wiens and M. Moss, eds., Issues and perspectives in landscape ecology, pp. 112–119. Cambridge: Cambridge University Press.

McGarigal, K., S.A. Cushman, M.C. Neel, and E. Ene. 2002. FRAGSTATS: Spatial pattern analysis program for categorical maps. Computer software program. University of Massachusetts, Amherst. http://www.umass.edu/landeco/research/fragstats/fragstats.html.

McGarigal, K., S.A. Cushman, M.C. Neel, and E. Ene. 2008. Habitat fragmentation. *In* K.K. Kemp, ed., Encyclopedia of geographic information science, p. 149–151. Thousand Oaks, CA: SAGE Publications.

McGarigal, K., S. Tagil, and S.A. Cushman. 2009. Surface metrics: An alternative to patch metrics for the quantification of landscape structure. Landscape Ecology 24:433–450. doi:10.1007/s10980-009-9327-y.

McGowan, K., and K. Corwin. 2008. The second atlas of breeding birds in New York State. Ithaca, NY: Cornell University Press.

McGowan, K., and B. Zuckerberg. 2008. Summary of results. *In* K. McGowan and K. Corwin, eds., The second atlas of breeding birds in New York State, pp. 15–42. Ithaca, NY: Cornell University Press.

McGuire, M., A. Gangopadhyay, L. Smith-Lovin, and J.M. Cook. 2008. A user-centered design for a spatial data warehouse for data exploration in environmental research. Ecological Informatics 3:273–285.

McPherson, M., L. Smith-Lovin, and J.M. Cook. 2001. Birds of a feather: Homophily in social networks. Annual Review of Sociology 27:415–444.

Means, B. 1998. Melding authentic science, technology, and inquiry-based teaching: Experiences of the GLOBE program. Journal of Science Education and Technology 7:97–105.

Merriam, S.B., R.S. Caffarella, and L.M. Baumgartner. 2006. Learning in adulthood: A comprehensive guide, 3rd ed. San Francisco: Jossey-Bass.

Michaelidou, M.J., D. Decker, and J. Lassoie. 2002. The interdependence of ecosystem and community viability: A theoretical framework to guide research and application. Society & Natural Resources 15:599–616.

Michener, W.K. 2006. Meta-information concepts for ecological data management. Ecological Informatics 1:3–7.

Miller, E.K., A. Vanarsdale, G.J. Keeler, A. Chalmers, L. Poissant, N.C. Kamman, and R. Brulotte. 2005. Estimation and mapping of wet and dry mercury deposition across northeastern North America. Ecotoxicology 14:53–70.

Miller, J. 2006. Civic scientific literacy in Europe and the United States. Montreal: World Association for Public Opinion Research.

Miller, J.D. 1983. Scientific literacy: A conceptual and empirical review. Daedalus 112:29–48.

Miller, J.D. 1998. The measurement of civic scientific literacy. Public Understanding of Science 7:203–223.

Miller, J.D. 2004. Public understanding of, and attitudes toward, scientific research: What we know and what we need to know. Public Understanding of Science 13:273–294.

Miller-Rushing, A.J., and R.B. Primack. 2008. Global warming and flowering times in Thoreau's Concord: A community perspective. Ecology 89:332–341.

Milwright, R.D.P. 2006. Post-breeding dispersal, breeding site fidelity and migration/wintering areas of migratory populations of Song Thrush *Turdus philomelos* in the Western Palearctic. Ringing & Migration 23:21–32.

Mitchell, T. 1997. Machine learning. New York: WCB/McGraw-Hill.

Moreau, R.E. 1944. Clutch size: A comparative study, with reference to African birds. Ibis 86:286–347.

Moss, D.M., E.D. Abrams, and J.A. Kull. 1998. Can we be scientists too? Secondary students' perceptions of scientific research from a project-based classroom. Journal of Science Education and Technology 7:149–161.

Munson, M.A., K. Webb, D. Sheldon, D.F. Fink, W.M. Hochachka, M.I. Iliff, M. Riedewald, D. Sorokina, B.L. Sullivan, C.L. Wood, et al. 2009. The eBird reference dataset, http://www.avianknowledge.net (accessed 4 July 2011). Ithaca, NY: Cornell Lab of Ornithology and National Audubon Society.

Murphy, A. 1998. Students and scientists take a "Lichen" to air quality assessment in Ireland. Journal of Science Education and Technology 7:107–113.

NADP NTN. 2001. Deposition data by site. Champaign, IL: NADP Program Office Illinois State Water Survey.

Nasir, N.S., A.S. Rosebery, B. Warren, and C.D. Lee. 2006. Learning as a cultural process: Achieving equity through diversity. *In* R.K. Sawyer, ed., The Cambridge handbook of the learning sciences, pp. 489–504. New York: Cambridge University Press.

National Research Council. 1996. National science education standards. Washington, DC: National Academies Press.

National Research Council. 1997. Taking science to school: Learning and teaching science in grades K–8. R.A. Duschl, H.A. Schweingruber, and A.W. Shouse, eds. Washington, DC: National Academies Press.

National Research Council. 2000. Inquiry and the national science education standards. Washington, DC: National Academy Press.

National Research Council. 2002. Scientific research in education. Washington, DC: National Academy Press.

National Research Council. 2007. Taking science to school. Washington, D.C., National Academies Press.

National Research Council. 2009. Learning science in informal environments: People, places, and pursuits. Washington, DC: National Academies Press.

National Science Board. 1996. Science and technology: Public attitudes and public understanding. *In* Science and engineering indicators: 1996, Chapter 7. Washington, DC: U.S. Government Printing Office.

National Science Board. 2002. Science and technology: Public attitudes and public understanding. *In* Science and engineering indicators: 2002 Chapter 7. Washington, DC: U.S. Government Printing Office.

National Science Foundation. 2009. A lifelong love of learning, http://www.nsf.gov/about/history/nsf0050/education/lifelonglove.htm, (accessed 4 July 2011). Washington, DC: National Science Foundation.

Nerbonne, J.F., and K.C. Nelson. 2004. Volunteer macroinvertebrate monitoring in the United States: Resource mobilization and comparative state structures. Society and Natural Resources 17:817–839.

Nerbonne, J.F., and K.C. Nelson. 2008. Volunteer macroinvertebrate monitoring: Tensions among group goals, data quality, and outcomes. Environmental Management 42:470.

Newhouse, M.J., P.P. Marra, and L.S. Johnson. 2008. Reproductive success of House Wrens in suburban and rural landscapes. Wilson Journal of Ornithology 120:99–104.

NEPAD. 2005. African post-conflict reconstruction policy framework, http://www.africanreview.org/docs/conflict/PCR%20Policy%20Framework.pdf (accessed 12 July 2011). Johannesburg, South Africa: Secretariat—Governance, Peace and Security Programme: New Partnership for Africa's Development.

Newson, S.E., K.L. Evans, D.G. Noble, J.J.D. Greenwood, and K.J. Gaston. 2008. Use of distance sampling to improve estimates of national population sizes for common and widespread breeding birds in the UK. Journal of Applied Ecology 45:1330–1338. doi:10.1111/j.1365–2664.2008.01480.x.

Newson, S.E., D.I. Leech, C.M. Hewson, H.Q.P. Crick, and P.V. Grice. 2010. Potential impact of grey squirrels *Sciurus carolinensis* on woodland bird populations in England. Journal of Ornithology 151:211–218. doi:10.1007/s10336–009–0445–8.

Newson, S.E., N. Ockendon, A.C. Joys, D.G. Noble, and S.R. Baillie. 2009. Comparison of habitat-specific trends in the abundance of breeding birds in the UK. Bird Study 56:233–243.

Newsweek. 2009. Toxic tsunami. 18 July. http://www.newsweek.com/id/207445/page/1 (accessed 28 August 2009).

Nice, M.M. 1937. Studies in the life history of the song sparrow. Vol. 1: A population study of the song sparrow. New York: Linnaean Society of New York.

Nice, M.M. 1943. Studies in the life history of the song sparrow. Vol. 2: The behavior of the song sparrow and other passerines. New York: Linnaean Society of New York.

Nickerson, R.S. 1998. Confirmation bias: A ubiquitous phenomenon in many guises. Review of General Psychology 2:175–220.

Nisbet, E., J. Zelenski, and S. Murphy. 2009. The nature relatedness scale. Environment and Behavior 41:715–740.

North American Bird Conservation Initiative, U.S. Committee. 2009. The state of the birds: United States of America. U.S. Department of Interior, Washington, DC.

North American Bird Conservation Initiative, U.S. Committee. 2010. The state of the birds: 2010 report on climate change. U.S. Department of Interior, Washington, DC.

Oberhauser, K.S., I. Gebhard, C. Cameron, and S. Oberhauser. 2007. Parasitism of monarch butterflies (*Danaus plexippus*) by *Lespesia archippivora* (Diptera: Tachinidae). American Midland Naturalist 157:312–328.

Oberhauser, K.S., M.D. Prysby, H.R. Mattila, D.E. Stanley-Horn, M.K. Sears, G. Dively, E. Olson, J.M. Pleasants, W.-K.F. Lam, and R. Hellmich. 2001. Temporal and spatial overlap between monarch larvae and corn pollen. Proceedings of the National Academy of Sciences 98:11913–11918.

Ockendon, N., S.E. Davis, T. Miyar, and M.P. Toms. 2009. Urbanization and time of arrival of common birds at garden feeding stations. Bird Study 56:405–410.

Ockendon, N., S.E. Davis, M.P. Toms, and S. Mukherjee. 2009. Eye size and the time of arrival of birds at garden feeding stations in winter. Journal of Ornithology 150:903–908. doi:10.1007/s10336–009–0412–4.

O'Connor, R.J. 1980a. Pattern and process in great tit (Parus major) populations in Britain. Ardea 68:165–183.

O'Connor, R.J. 1980b. Population regulation in the Yellowhammer *Emberiza citrinella* in Britain. *In* H. Oelke, ed., Proceedings of the Sixth International Conference on Bird Census Work and Fourth Meeting of the European Ornithological Atlas Committee: Bird census work and nature conservation, pp. 190–200. Lengede: Dachverband Deutscher Avifaunisten.

O'Connor, R.J. 1985. Behavioural regulation of bird populations: A review of habitat use in relation to migration and residency. *In* R.M. Sibley and R.H. Smith, eds., Behavioural ecology: Ecological consequences of adaptive behaviour, pp. 105–142. Oxford: Blackwell Scientific Publications.

O'Connor, R.J., 1986. Dynamical aspects of avian habitat use. *In* J. Verner, M.L. Morrison, and C.J. Ralph, eds., Wildlife 2000: Modelling habitat relationships of terrestrial vertebrates, pp. 235–240. Madison: University of Wisconsin.

O'Connor, R.J., and R.J. Fuller. 1985. Bird census and atlas studies: Bird population responses to habitat. *In* K. Taylor, R.J. Fuller, and P.C. Lack, eds., Sixth International Conference on Bird Census Work: Bird censuses and atlas studies, pp. 197–211. Tring: British Trust for Ornithology.

Oggier, P., S. Zschokke, and B. Baur. 1998. A comparison of three methods for assessing the gastropod community in dry grasslands. Pedobiologia 42:348–357.

Oliver, P., G. Marwell, and R. Teixeira. 1985. A theory of the critical mass. Part 1: Interdependence, group heterogeneity, and the production of collective action. American Journal of Sociology 91:522–556.

Olson, M., Jr. 1965. The logic of collective action: Public goods and the theory of groups. Cambridge, MA: Harvard University Press.

Ostrom, E. 1998. A behavioral approach to the rational choice theory of collective action: Presidential address, American Political Science Association, 1997. American Political Science Review 92:1–22.

Overdevest, C., C. Huyck Orr, and K. Stepenuck. 2004. Volunteer stream monitoring and local participation in natural resource issues. Research in Human Ecology 11:177–185.

Pahl-Wostl, C., M. Craps, A. Dewulf, E. Mostert, D. Tabara, and T. Tailleu. 2007. Social learning and water resources management. Ecology and Society 12:5.

Palmberg, I.E., and J. Kuru. 2000. Outdoor activities as a basis for environmental responsibility. Journal of Environmental Education 31:32–36.

Palmer, J.A. 1993. Development of concern for the environment and formative experiences of educators. Journal of Environmental Education 24:26–30.

Palmer, J.A. and J. Suggate, 1996. Influences and experiences affect pro-environment behavior of educators. Environmental Education Research 2:109–121.

Palmer, J.A., J. Suggate, I. Robottom, and P. Hart. 1999. Significant life experiences and formative infuences on the development of adults' environmental awareness in the UK, Australia and Canada. Environmental Education Research 5:181–200.

Paradis, E., S. Baillie, W. Sutherland, and R. Gregory. 2002. Dispersal and spatial scale affect synchrony in spatial population dynamics. Ecology Letters 2:114–120.

Paradis, E., S.R. Baillie, W.J. Sutherland, and R.D. Gregory. 1998. Patterns of natal and breeding dispersal in birds. Journal of Animal Ecology 67:518–536.

Park, J., and M. Schaller. 2009. Does attitude similarity serve as a heuristic cue for kinship? Evidence of an implicit cognitive association. Evolution and Human Behavior 26:158–170.

Parker, J.D., and M.H. McDonough. 1999. Environmentalism of African Americans: An analysis of the barriers and subculture theories. Environment and Behavior 31:155–177.

Parrish, J.K., N. Bond, H. Nevins, N. Mantua, R. Loeffel, W.T. Peterson, and J.T. Harvey. 2007. Beached birds and physical forcing in the California Current System. Marine Ecology Progress Series 352:275–288. doi:10.3354/meps07077.

Pattengill-Semmens, C.V., and B.X. Semmens. 2003. Conservation and management applications of the reef volunteer fish monitoring program. Environmental Monitoring and Assessment 81:43–50.

Peach, W.J., S.R. Baillie, and L. Underhill. 1991. Survival of British Sedge Warblers Acrocephalus schoenobaenus in relation to West African rainfall. Ibis 133:300–305.

Peakall, D.B. 1970. The eastern bluebird: Its breeding season, clutch size, and nesting success. Living Bird 9:239–255.

Pearl, R.E. 1974. The present status of science attitude measurement: History, theory and availability of measurement instruments. School Science and Mathematics 74:375–381.

Pearson, R.G. 2007. Species distribution modeling for conservation educators and practitioners. Center for Biodiversity and Conservation, American Museum of Natural History, http://ncep.amnh.org/ (accessed 4 July 2011). New York: American Museum of Natural History.

Peeters, G., and J. Czapinski. 1990. Positive-negative asymmetry in evaluations: The distinctions between affective and informational negativity effects. European review of social psychology 1:33–60.

Pelling, M. 2007. Learning from others: The scope and challenges for participatory disaster risk assessment. Disasters 31:373–385.

Penuel, W.R., M. Bienkowski, L. Gallagher, C. Korbak, W. Sussex, R. Yamaguchi, and B.J. Fishman. 2006. GLOBE year 10 evaluation: Into the next generation. SRI International. http://www.globe.gov/fsl/evals/y10full.pdf.

Penuel, W.R., B.J. Fishman, R. Yamaguchi, and L.P. Gallagher. 2007. What makes professional development effective? Strategies that foster curriculum implementation. American Educational Research Journal 44:921–958.

Penuel, W.R., and B. Means. 2004. Implementation variation and fidelity in an inquiry science program: An analysis of GLOBE data reporting patterns. Journal of Research in Science Teaching 41:294–315.

Penuel, W.R., L. Shear, C. Korbak, and E. Sparrow. 2005. The roles of regional partners in supporting an international earth science education program. Science Education 89:956–979.

Petersen, A., A. Anderson, C. Wilkinson, and A. Stuart. 2007. Nanotechnology, risk and society. Health, Risk & Society 9:117–124.

Peterson, N.J., and H.R. Hungerford. 1981. Developmental variables affecting environmental sensitivity in professional environmental educators. *In* A.B. Sacks, L.A. Iozzi, J.M. Schultz, and R. Wilke, eds., Current issues in environmental education and environmental studies. Selected Papers from the Annual Conference of the National Association for Environmental Education (10th, Gilbertsville, KY) 7:111–114. Columbus, OH: ERIC.

Phillips, T., B.V. Lewenstein, and R. Bonney. 2006. A case study of citizen science. *In* C. Donghong, J. Metcalfe, and B. Schiele, eds., At the human scale: International practices in science communication, pp. 317–334. Beijing: Science Press.

Popper, K.R. 1959. The logic of scientific discovery. London: Hutchinson.

Postmes, T., and S. Brunsting. 2002. Collective action in the age of the Internet: Mass communication and online mobilization. Social Science Computer Review 20:290–301.

Poulin, R.G., and R.M. Brigham. 2001. Effects of supplemental calcium on the growth rate of an insectivorous bird, the purple martin (*Progne subis*). Ecoscience 8:151–156.

Powell, M., and M. Colin. 2008. Meaningful citizen engagement in science and technology: What would it really take? Science Communication 30:126–136.

Powell, M., and M. Colin. 2009. Participatory paradoxes: Facilitating citizen engagement in science and technology from the top down? Bulletin of Science, Technology, and Society 29:325–342.

Prasad, A., L. Iverson, and A. Liaw. 2006. Newer classification and regression tree techniques: Bagging and random forests for ecological prediction. Ecosystems 190:231–259.

Prater, A.J. 1981. Estuary birds of Britain and Ireland. Calton: T. & A.D. Poyser.

Primack, R.B., H. Higuchi, and A.J. Miller-Rushing. 2009. The impact of climate change on cherry trees and other species in Japan. Biological Conservation 142:1943–1949. doi:10.1016/j.biocon.2009.03.016.

Primack, R.B., A.J. Miller-Rushing, and K. Dharaneeswaran. 2009. Changes in the flora of Thoreau's Concord. Biological Conservation 142:500–508. doi10.1016/j.biocon.2008.10.038.

Prysby, M., and K. Oberhauser. 1999. Large-scale monitoring of monarch populations. *In* Proceedings of the North American Conference on the Monarch Butterfly, pp. 379–384. Montreal: Commission for Environmental Cooperation.

Prysby, M., and K. Oberhauser. 2004. Temporal and geographical variation in monarch densities: Citizen scientists document monarch population patterns. *In* K.S. Oberhauser and M.J. Solensky, eds., The monarch butterfly: Biology and conservation, pp. 9–20. Ithaca, NY: Cornell University Press.

Prysby, M.D. 2004. Natural enemies and survival of monarch eggs and larvae. *In* K.S. Oberhauser and M.J. Solensky, eds., The monarch butterfly: Biology and conservation, pp. 27–38. Ithaca, NY: Cornell University Press.

Rabosky, D.L., and I.J. Lovette. 2008. Explosive evolutionary radiations: Decreasing speciation or increasing extinction through time? Evolution 62:1866–1875. doi:10.1111/j.1558-5646.2008.00409.x.

Rahm, J. 2002. Emergent learning opportunities in an inner-city youth gardening program. Journal of Research in Science Teaching 39:164–184.

Ravetz, I.R. 1999. Papers: What is post-normal science? Futures 31:647–654.

Rea, B., K. Oberhauser, and M. Quinn. 2002. A field guide to invertebrates on milkweed. Union, WV: Bas Relief Publishing Group.

Rich, T.D., C.J. Beardmore, H. Berlanga, P.J. Blancher, M.S.W. Bradstreet, G.S. Butcher, D.W. Demarest, E.H. Dunn, W.C. Hunter, E.E. Inigo-Elias, et al. 2004. Partners in flight North American landbird conservation plan. Ithaca, NY: Cornell Lab of Ornithology.

Rickinson, M. 2001. Learners and learning in environmental education: A critical review of the evidence. Environmental Education Research 7:207–320.

Ricklefs, R.E. 2000. Lack, Skutch, and Moreau: The early development of life-history thinking. Condor 102:3–8.

Rimmer, C.C., K.P. McFarland, D.C. Evers, E.K. Miller, Y. Aubry, D. Busby, and R.J. Taylor. 2005. Mercury concentrations in Bicknell's thrush and other insectivorous passerines in Montane forests of northeastern North America. Ecotoxicology 14:223–240.

Risely, K., D. Noble, and S. Baillie. 2009. The breeding bird survey 2008. BTO research report 537, British Trust for Ornithology, Thetford, http://www.bto.org/birdtrends.

Robbins, C., S. Droege, M. Perry, J. Lowe, and R. James. 2006. Historical bird monitoring data, http://www.pwrc.usgs.gov/birds/histdata.html (accessed 4 July 2011). U.S. Geological Survey, Patuxent Wildlife Research Center.

Robin, J., E. Levine, and S. Riha. 2005. Utilizing satellite imagery and GLOBE student data to model soil dynamics. Ecological Modelling 185:133–145.

Robinson, J.C. 2005. Relative prevalence of African Americans among bird watchers. USDA Forest Service Gen. Tech. Rep. PSW-GTR-191–2005.

Robinson, R.A., D.E. Balmer, and J.H. Marchant. 2008. Survival rates of hirundines in relation to British and African rainfall. Ringing & Migration 24:1–6.

Robinson, S.K., and W.D. Robinson. 2001. Avian nesting success in a selectively harvested north temperate deciduous forest. Conservation Biology 15:1763–1771.

Rock, B.N., and G.N. Lauten. 1996. K–12th grade students as active contributors to research investigations. Journal of Science Education and Technology 5:255–266.

Rodriguez, J., F. Vos, R. Below, and D. Guha-Sapir. 2009. Annual disaster statistical review 2008: The numbers and trends. Brussels: Centre for Research on the Epidemiology of Disasters.

Rogers, E. 2003. Diffusion of innovations, 5th ed. New York: Free Press.

Root, T.L. 1988. Environmental factors associated with avian distributional boundaries. Journal of Biogeography 15:489–505.

Rosenberg, K.V., R.S. Hames, R.W.J. Rohrbaugh, S. Barker Swarthout, J.D. Lowe, and A.A. Dhondt. 1999. A land manager's guide to improving habitat for scarlet tanagers and other forest interior birds. Ithaca, NY: Cornell Lab of Ornithology.

Rosenberg, K.V., R.S. Hames, R.W.J. Rohrbaugh, S.B. Swarthout, J.D. Lowe, and A.A. Dhondt. 2003. A land manager's guide to improving habitat for forest thrushes. Ithaca, NY: Cornell Lab of Ornithology.

Rosenberg, K.V., J.D. Lowe, and A.A. Dhondt. 1999. Effects of forest fragmentation on breeding tanagers: A continental perspective. Conservation Biology 13:568–583.

Rossi, H.P., M.W. Lipsey, and H.E. Freeman, 2004. Evaluation: A systematic approach, 7th ed. Thousand Oaks, CA: SAGE Publications.

Roth, W.M., and S. Lee. 2002. Scientific literacy as collective praxis. Public Understanding of Science 11:33–56.

Roux, K.E., and P.P. Marra. 2007. The presence and impact of environmental lead in passerine birds along an urban to rural land use gradient. Archives of Environmental Contamination and Toxicology 53:261–268. doi:10.1007/s00244–006–0174–4.

Rowe, G., and L.J. Frewer. 2005. A typology of public engagement mechanisms. Science, Technology, & Human Values 30:251–290.

Roy, D.B., and T.H. Sparks. 2000. Phenology of British butterflies and climate change. Global Change Biology 6:407–416.

Royle, J.A., M. Kéry, R. Gautier, and H. Schmid. 2007. Hierarchical spatial models of abundance and occurrence from imperfect survey data. Ecological Monographs 77:465–481.

Ruitenbeek, J., and C. Cartier. 2001. The invisible wand: Adaptive co-management as an emergent strategy in complex bio-economic systems. Occasional paper, Center for International Forestry Research, Bogor, Indonesia.

Rutherford, F.J., and A. Ahlgren. 1989. Science for all Americans. New York: Oxford University Press.

Ryder, T.B., R. Reitsma, B. Evans, and P.P. Marra. 2010. Quantifying avian nest survival along an urbanization gradient using citizens and scientist generated data. Ecological Applications 20:419–426.

Sackett, D.L. 1979. Bias in analytic research. Journal of Chronic Disease 32:51–63.

Sadler, T.D., and D.L. Zeidler. 2005. The significance of content knowledge for informal reasoning regarding socioscientific issues: Applying genetics knowledge to genetic engineering issues. Science Education 89:71–93.

Santos, F.C., M.D. Santos, and J.M. Pacheco. 2008. Social diversity promotes the emergence of cooperation in public goods games. Nature 454:213–216.

Sauer, J.R., J.E. Hines, and J. Fallon. 2004. The North American breeding bird survey results and analysis 1966–2003, 1st ed. Laurel, MD: Patuxent Wildlife Research Center.

Sauer, J.R., and W.A. Link. 2002. Hierarchical modeling of population stability and species group attributes from survey data. Ecology 83:1743–1751.

Saunders, D.A., R.J. Hobbs, and C.R. Margules. 1991. Biological consequences of ecosystem fragmentation: A review. Conservation Biology 5:18–32.

Saunders, W.L., and D.H. Dickinson. 1979. A comparison of community college students' achievement and attitude changes in a lecture-only and lecture-laboratory approach to general education biological science courses. Journal of Research in Science Teaching 16:459–464.

Scheuhammer, A.M. 1991. Effects of acidification on the availability of toxic metals and calcium to wild birds and mammals. Environmental Pollution 71:329–375.

Scheuhammer, A.M. 1996. Influence of reduced dietary calcium on the accumulation and effects of lead, cadmium, and aluminum in birds. Environmental Pollution 94:337–343.

Schmeller, D.S., P.-Y. Henry, R. Julliard, B. Gruber, J. Clobert, F. Dziock, S. Lengyel, P. Nowicki, E. Déri, E. Budrys, et al. 2009. Advantages of volunteer-based biodiversity monitoring in Europe. Conservation Biology 23:307–316. doi:10.1111/j.1523–1739.2008.01125.x

Schwartz, D.L., J.D. Bransford, and D. Sears. 2005. Efficiency and innovation in transfer. In J. Mestre, ed., Transfer of learning: Research and perspectives, 1–51. Boston: Information Age Publishing.

Schwartz, M.D., R. Ahas, and A. Aasa. 2006. Onset of spring starting earlier across the Northern Hemisphere. Global Change Biology 12:343–351. doi:10.1111/j.1365–2486.2005.01097.x.

Schwartz, S.H. 1977. Normative influences on altruism. Advances in Experimental Social Psychology 10:221–279.

Scott, J. 2000. Social network analysis: A handbook, 2nd ed. Thousand Oaks, CA: Sage Publications.

Scott, J. 2002. Predicting species occurrences: Issues of accuracy and scale. Washington, DC: Island Press.

Shamos, M.H. 1995. The myth of scientific literacy. New Brunswick, NJ: Rutgers University Press.

Sharrock, J. 1976. The atlas of breeding birds in Britain and Ireland. Tring: British Trust for Ornithology.

Shirk, J.L. 2004. Student-scientist partnerships: Student inquiry and data accuracy in a salamander monitoring project. Master's thesis], Cornell University, Ithaca, NY.

Sibthorp, J., K. Paisley, and J. Gookin. 2007. Exploring participant development through adventure-based programming: A model from the national outdoor leadership school. Leisure Sciences 29:1–18.

Sinclair, K.D., and B.A. Knuth. 2000. Nonindustrial private forest landowner use of geographic data: A precondition for ecosystem-based management. Society & Natural Resources 13:521–536.

Siriwardena, G.M., S.R. Baillie, S.T. Buckland, R.M. Fewster, J.H. Marchant, and J.D. Wilson. 1998. Trends in the abundance of farmland birds: A quantitative comparison of smoothed Common Birds Census indices. Journal of Applied Ecology 35:24–43.

Siriwardena, G.M., S.R. Baillie, H.Q.P. Crick, and J.D. Wilson. 2000. The importance of variation in the breeding performance of seed-eating birds in determining their population trends on farmland. Journal of Applied Ecology 37:128–148. doi:10.1046/j.1365–2664.2000.00484.x.

Siriwardena, G.M., S.R. Baillie, and J.D. Wilson. 1998. Variation in the survival rates of some British passerines with respect to their population trends on farmland. Bird Study 45:276–292. doi:10.1080/00063659809461099.

Skelly, S.M., and J.M. Zajiek. 1998. The effect of an interdisciplinary garden program on the environmental attitudes of elementary students. HortTechnology 8:579–583.

Skutch, A.F. 1949. Do tropical birds rear as many young as they can nourish? Ibis 91:430–455.

Slovic, P. 1987. Perception of risk. Science 236:280–285.

Smith, C.R. 1990. Handbook for atlasing North American breeding birds. Vermont Institute of Natural Science, Woodstock, for the North American Ornithological Atlas Committee.

Smith, F., and C. Hausafus. 1998. Relationship of family support and ethnic minority students' achievement in science and mathematics. Science Education 82:111–129.

Smith, N.K., J.T. Larsen, T.L. Chartrand, J.T. Cacioppo, H.A. Katafiasz, and K.E. Moran. 2006. Being bad isn't always good: Affective context moderates the attention bias toward negative information. Journal of Personality and Social Psychology 90:210–220.

Smith-Sebasto, N.J., and V.L. Obenchain. 2009. Students' perceptions of the residential environmental education program at the New Jersey School of Conservation. Journal of Environmental Education 40:50–62.

Snow, D.W. 1966. Population dynamics of the Blackbird. Nature 211:1231–1233.

Snow, L.C., L.C. Newson, A.J. Musgrove, P.A. Cranswick, H.Q.P. Crick, and J.W. Wilesmith. 2007. Risk-based surveillance for H5N1 avian influenza virus in wild birds in Great Britain. Veterinary Record 161:775–781.

Snyder, D., and S. Kaufman. 2004. An overview of nonindigenous plant species in New Jersey. Department of Environmental Protection, Trenton, NJ.

Sobel, D. 2004. Place-based education: Connecting classrooms and communities. Great Barrington, MA: Orion Society.

Sorokina, D., R. Caruana, M. Riedewald, and D. Fink. 2008. Detecting statistical interactions with additive groves of trees. Proceedings of the 25th International Conference on Machine Learning (ICML 2008), 5–9 July 2008, Helsinki, pp. 1000–1007. New York: Omnipress.

Sorokina, D., R. Caruana, M. Riedewald, W.M. Hochachka, and S. Kelling. 2009. Detecting and interpreting variable interactions in observational ornithology data. In Proceedings of the International Conference on Domain Driven Data Mining Workshops (DDDM), pp. 64–69. Miami, FL. Los Alamitos, CA: Conference Publishing Services. http://www.computer.org/portal/web/csdl/doi/10.1109/ICDMW.2009.84 (accessed 4 July 2011).

Spencer, R. 1983. The Trust from 1951 to 1982. In R. Hickling, ed., Enjoying ornithology, pp. 29–53. Tring: British Trust for Ornithology.

Stern, M.J., R.B. Powell, and N.M. Ardoin. 2008. What difference does it make? Assessing outcomes from participation in a residential environmental education program. Journal of Environmental Education 39:31–43.

Stern, P. 2000. Toward a coherent theory of environmentally significant behavior. Journal of Social Issues 56:407–424.

Strang, D.S., and S.A. Soule. 1998. Diffusion in organizations and social movements: From hybrid corn to poison pills. Annual Review of Sociology 24:265–290.

Stukas, A., K. Worth, E. Clary, and M. Snyder. 2009. The matching of motivations to affordances in the volunteer environment. Nonprofit and Voluntary Sector Quarterly 38:5–28.

Sturgis, P., and N. Allum. 2004. Science in society: Re-evaluating the deficit model of public attitudes. Public Understanding of Science 13:55–74.

Sullivan, B., C. Wood, M. Iliff, R. Bonney, D. Fink, and S. Kelling. 2009. eBird: A citizen-based bird observation network in the biological sciences. Biological Conservation 142:2282–2292.

Sutcliffe, S., and M. Bradstreet. 1994. Agreement between Cornell Laboratory of Ornithology and Long Point Bird Observatory concerning management of Project FeederWatch. Ithaca, NY, and Port Rowan, ON.

Sward, L.L. 1999. Significant life experiences affecting the environmental sensitivity of El Salvadoran environmental professionals. Environmental Education Research 5:201–206.

Swink F., and G. Wilhelm. 1994. Plants of the Chicago region, 4th ed. Indianapolis: Indiana Academy of Science.

Takashi, N. 2004. Spatial hierarchical approach in community ecology: A way beyond high context-dependency and low predictability in local phenomena. Population Ecology 46:105.

Tanner C., 1999. Constraints on environmental behavior. Journal of Environmental Psychology 19:145–157.

Tanner T., 1980. Significant life experiences: A new research area in environmental education. Journal of Environmental Education 11:20–24.

Tanner, W.P., and J.A. Swets. 1954. A decision-making theory of visual detection. Psychological Review 61:6.

Tear, T.H., P. Kareiva, P.L. Angermeier, P. Comer, B. Czech, R. Kautz, L. Landon, D. Mehlman, K. Murphy, M. Ruckelshaus, et al. 2005. How much is enough? The recurrent problem of setting measurable objectives in conservation. Bioscience 55:835–849.

Thogmartin, W.E., J.R. Sauer, and M.G. Knutson. 2004. A hierarchical spatial model of avian abundance with application to Cerulean Warblers. Ecological Applications 14:1766–1779.

Thomas, C.D., and J.J. Lennon. 1999. Birds extend their ranges northwards. Nature 399:213.

Thompson, W.L. 2004. Sampling rare or elusive species: Concepts, designs, and techniques for estimating population parameters. Washington, DC: Island Press.

Tidball, K.G., and M.E. Krasny. 2007. From risk to resilience: What role for community greening and civic ecology in cities? In A.E.J. Wals, ed., Social learning towards a more sustainable world, pp. 149–164. Wagengingen, Netherlands: Wagengingen Academic Press.

Tidball, K.G., and M.E. Krasny. 2008a. "Raising" urban resilience: Community forestry and greening in cities post-disaster/conflict. Resilience, adaptation and transformation in turbulent times, 14–18 April, Stockholm: Resilience Alliance Conference.

Tidball, K.G., and M.E. Krasny. 2008b. Trees and rebirth: Urban community forestry in post-Katrina resilience. CFERF report, Berkeley, CA: Community Forestry and Environmental Research Fellows Program.

Tidball, K.G., E.D. Weinstein, S. Kaisler, R. Grossman-Vermaas, and S. Tousley. 2008. Stake-holder asset-based planning environment. Report, Logos Technologies, Cornell University, and International Sustainable Systems. http://www.sci-links.com/files/SHAPE_FINAL_REPORT_web.pdf (accessed 12 July 2011).

Tilgar, V., R. Mand, and M. Magi. 2002. Calcium shortage as a constraint on reproduction in great tits *Parus major*: A field experiment. Journal of Avian Biology 33:407–413.

Titeux, N., M. Dufrene, J.P. Jacob, M. Paquay, and P. Defourny. 2004. Multivariate analysis of a fine-scale breeding bird atlas using a geographical information system and partial canonical correspondence analysis: Environmental and spatial effects. Journal of Biogeography 31:1841–1856.

Tomasek, T.M. 2006. Student cognition and motivation during the Classroom BirdWatch citizen science project. PhD dissertation, University of North Carolina at Greensboro.

Toms, M.P., and P. Sterry. 2008. Garden birds and wildlife. Thetford: British Trust for Ornithology.

Trautmann, N.M. 2009. Interactive learning through web-mediated peer review of student science reports. Educational Technology Research & Development 57:685–704. doi:10.1007/s11423–007–9077-y.

Trautmann, N.M., and J.G. MaKinster. 2005. Teacher/scientist partnerships as professional development: Understanding how collaboration can lead to inquiry. AETS International Conference, 19–23 January 2005; Colorado Springs, CO: Association for the Education of Teachers in Science.

Trochim, W.K. 2006. Research methods knowledge base, http://www.socialresearchmethods.net/kb/intreval.php (accessed 4 July 2011). Ithaca, NY: Web Center for Social Research Methods.

Trumbull, D.J., R. Bonney, D. Bascom, and A. Cabral. 2000. Thinking scientifically during participation in a citizen science project. Science Education 84:254–275.

Trumbull, D.J., R. Bonney, and N. Grudens-Schuck. 2005. Developing materials to promote inquiry: Lessons learned. Science Education 89:879–900.

Trumbull, D.J., G. Scarano, and R. Bonney. 2006. Relations among two teachers' practices and beliefs, conceptualizations of the nature of science, and their implementation of student independent inquiry projects. International Journal of Science Education 28:1717–1750.

Trzcinski, M.K., L. Fahrig, and G. Merriam. 1999. Independent effects of forest cover and fragmentation on the distribution of forest breeding birds. Ecological Applications 9:586–593.

Tudor, M.T., and K.M. Dvornich. 2001. The NatureMapping program: Resource agency environmental reform. Journal of Environmental Education 12:8–14.

Tukey, J.W. 1977. Exploratory data analysis. Reading, MA: Addison-Wesley.

Turner, M.G. 1989. Landscape ecology: The effect of pattern on process. Annual Review of Ecology and Systematics 20:171–197.

Turner, M.G., R.H. Gardner, and R.V. O'Neill. 2001. Landscape ecology in theory and practice: Pattern and process. New York: Springer.

Tversky, A., and D. Kahneman. 1981. The framing of decisions and the psychology of choice. Science 211:453–458.

United Nations and World Bank. 2004. Practical guide to multilateral needs assessments in post-conflict situations. Social Development Papers no. 15. Washinton DC: World Bank.

Uriarte, M., H.A. Ewing, V.T. Eviner, and K.C. Weathers. 2007. Constructing a broader and more inclusive value system in science. BioScience 57:71–78.

U.S. Department of Education. 2007. Report of the Academic Competitiveness Council. Washington D.C.: ED Pubs.

USDA/NRCS. 1994. State soil geographic (STATSGO) data base: Data use information. Washington, DC: USDA Natural Resources Conservation Service, National Soil Survey Center.

USFWS. 2002. A 2001 national survey of fishing, hunting and wildlife associated recreation. Department of the Interior.

USFWS. 2009. Addendum to the 2006 national survey of fishing, hunting, and wildlife-associated recreation. Department of the Interior.

Valente, T. 1996. Social network thresholds in the diffusion of innovations. Social Networks 18:69–89.

van der Merwe, M., J.S. Brown, and W.M. Jackson. 2005. The coexistence of fox (*Sciurus niger*) and gray (*S. caroliniensis*) squirrels in the Chicago metropolitan area. Urban Ecosystems 8:335–347.

Van Vugt, M., M. Snyder, T. Tyler, and A. Biel, eds. 2000. Cooperation in modern society: Promoting the welfare of communities, states and organizations. London: Routledge.

van Zee, E.H. 2000. Analysis of a student-generated inquiry discussion. International Journal of Science Education 22:115–142.

van Zee, E.H., M. Iwasyk, A. Kurose, D. Simpson, and J. Wild. 2001. Student and teacher questioning during conversations about science. Journal of Research in Science Teaching 38:159–190.

Venier, L.A., J. Pearce, J.E. McKee, D.W. McKenney, and G.J. Niemi. 2004. Climate and satellite-derived land cover for predicting breeding bird distribution in the Great Lakes Basin. Journal of Biogeography 31:315–331.

Verbyla, D.L. 2001. A test of detecting spring leaf flush within the Alaskan boreal forest using ERS-2 and Radarsat SAR data. International Journal of Remote Sensing 22:1159–1165.

Villard, M.A., M.K. Trzcinski, and G. Merriam. 1999. Fragmentation effects on forest birds: Relative influence of woodland cover and configuration on landscape occupancy. Conservation Biology 13:774–783.

Vining, J., and A. Ebreo. 2002. Emerging theoretical and methodological perspectives on conservation behavior. *In* R.B. Bechtel and A. Churchman, eds., Handbook of environmental psychology, 541–558. New York: Wiley.

Visser, M.E., F. Adriaensen, J.H. van Balen, J. Blondel, A.A. Dhondt, S. van Dongen, C. du Feu, E.V. Ivankina, A.B. Kerimov, J. de Laet, et al. 2003. Variable response to large-scale climate change in European *Parus* populations. Proceedings of the Royal Society of London, series B. doi:10.1098/rspb.2002.2244.

Voříšek, P., A. Klvaňova, S. Wotton, and R.D. Gregory, eds. 2008. A best practice guide for wild bird monitoring schemes. Prague: Czech Society for Ornithology Royal Society for the Protection of Birds.

Voss, M.A., and C.B. Cooper. 2010. Using a free online citizen science project to teach observation and quantification of animal behavior. American Biology Teacher 72:437–443.

Wagner, M.W., U.P. Kreuter, R.A. Kaiser, and R.N. Wilkins. 2007. Collective action and social capital of wildlife management associations. Journal of Wildlife Management 71:1729–1738.

Walker, B.H., and D. Salt. 2006. Resilience thinking: Sustaining ecosystems and people in a changing world. Washington, DC: Island Press.

Walther, J.B., and J.K. Burgoon. 1992. Relational communication in computer-mediated interaction. Human Communication Researcher 19:50–88.

Ward Thompson, C., P. Aspinall, and A. Montarzino. 2008. The childhood factor: Adult visits to green places and the significance of childhood experiences. Environment and Behavior 40:111–143.

Weare, C., W.E. Loges, and N. Oztas. 2007. Email effects on the structure of local associations: A social network analysis. Social Science Quarterly 88:222–243.

Weber, E.P. 2000. A new vanguard for the environment: Grass-roots ecosystem management as a new environmental movement. Society and Natural Resources 13:237–259.

Weinstein, E., and K.G. Tidball. 2007. Environment shaping: An alternative approach to development and aid. Journal of Intervention and Statebuilding 1 (1).

Wells, J., K. Rosenberg, D. Tessaglia, and A. Dhondt. 1996. Population cycles in the Varied Thrush (*Ixoreus naevius*). Canadian Journal of Zoology 74:2062–2069.

Wells, J.V., K.V. Rosenberg, E.H. Dunn, D.L. Tessaglia-Hymes, and A.A. Dhondt. 1998. Feeder counts as indicators of spatial and temporal variation in winter abundance of resident birds. Journal of Field Ornithology 69:577–586.

Wells, N.M. 2000. At home with nature: Effects of "greenness" on children's cognitive functioning. Environment and Behavior 32:775–795.

Wells, N.M., and G.W. Evans. 2003. Nearby nature: A buffer of life stress among rural children. Environment and Behavior 35:311–330.

Wells, N.M., and K.S. Lekies. 2006. Nature and the life course: Pathways from childhood nature experiences to adult environmentalism. Children, Youth, and Environment 16:1–24.

Wenger, E. 1998. Communities of practice: Learning, meaning, and identity. New York: Cambridge University Press.

Wernham, C.V., M.P. Toms, J.H. Marchant, J.A. Clark, G.M. Siriwardena, and S.R. Baillie, eds. 2002. The migration atlas: Movements of the birds of Britain and Ireland. London: T. & A.D. Poyser.

Westat, J.F. 2002. The 2002 user-friendly handbook for project evaluation. Directorate for Education & Human Resources, Division of Research, Evaluation and Communication, National Science Foundation, Arlington, VA, http://www.nsf.gov/pubs/2002/nsf02057/nsf02057_l.pdf.

White, B.Y., and J.R. Frederiksen. 1998. Inquiry, modeling, and metacognition: Making science accessible to all students. Cognition and Instruction 16:3–118.

Wieczorek, J. 2007. Darwin Core TDWG Task Group, http://www.tdwg.org/activities/darwincore. Berkeley: Museum of Vertebrate Zoology, University of California.

Wiens, J.A. 2007. Foundation papers in landscape ecology. New York: Columbia University Press.

Wikipedia. 2009. Social Media. http://en.wikipedia.org/wiki/Social_media (accessed 4 July 2011).

Wilcove, D.S., C.H. McLellan, and A.P. Dobson. 1986. Habitat fragmentation in the temperate zone. *In* M.E. Soulé, ed., Conservation biology, pp. 237–256. Sunderland, MA: Sinauer.

Wilderman, C.C., A. Barron, and L. Imgrund. 2004. From the field: A service provider's experience with two models for community science. Community-Based Collaborative Research Consortium Journal (Spring), http://www.cbcrc.org/journal.html.

Wilderman, C.C., A. Barron, and L. Imgrund. 2005. Top down or bottom up? ALLARM's experience with two operational models for community science. 4th National Water Quality Monitoring Council conference, 17–20 May. Chattanooga, TN: National Water Monitoring Council.

Williamson, K. 1969. Habitat preferences of the Wren on English farmland. Bird Study 16:53–59.

Wilson, C.D., J.A. Taylor, S.M. Kowalski, and J. Carlson. 2010. The relative effects and equity of inquiry-based and commonplace science teaching on students' knowledge, reasoning, and argumentation. Journal of Research in Science Teaching 47:276–301.

Wilson, J., A. Evans, and P. Grice. 2009. Bird conservation and agriculture. Cambridge: Cambridge University Press.

Windschitl, M., and H. Buttemer. 2000. What should the inquiry experience be for the learner? American Biology Teacher 62:146–351.

Winstanley, D., R. Spencer, and K. Williamson. 1974. Where have all the Whitethroats gone? Bird Study 21:1–14.

With, K.A. 2002. The landscape ecology of invasive spread. Conservation Biology 16:1192–1203.

W.K. Kellogg Foundation. 2004. Logic model development guide. Battle Creek, MI: W.K. Kellogg Foundation.

Wood, S.N. 2006. Generalized additive models: An introductionwith R. Boca Raton, FL: Chapman & Hall CRC.

Wu, J.G., and R. Hobbs. 2002. Key issues and research priorities in landscape ecology: An idiosyncratic synthesis. Landscape Ecology 17:355–365.

Wymer, W.W. 2003. Differentiating literacy volunteers: A segmentation analysis for target marketing. International Journal of Nonprofit and Voluntary Sector Marketing 8:267–285.

Yosso, T.J. 2005. Whose culture has capital? A critical race theory discussion of community cultural wealth. Race Ethnicity and Education 8:69–91.

Young, B.E. 1994. Geographic and seasonal patterns of clutch-size variation in house wrens. Auk 111:545–555.

Zelezny, L.C., P. Chua, and C. Aldrich. 2000. Elaborating on gender differences in environmentalism. Journal of Social Issues 56:443–457.

Zuckerberg, B., F. Huettmann, and J. Frair. 2011. Proper data management as a scientific foundation for reliable species distribution modeling. In C.A. Drew, Y.F. Wiersma and F. Huettmann, eds., Predictive species and habitat modeling in landscape ecology, 45–70. New York: Springer Verlag.

Zuckerberg, B., and W.F. Porter. 2010. Thresholds in the long-term responses of breeding birds to forest cover and fragmentation. Biological Conservation 143:952–962.

Zuckerberg, B., W.F. Porter, and K. Corwin. 2009. The consistency and stability of abundance-occupancy relationships in large-scale population dynamics. Journal of Animal Ecology 78:172–181. doi:10.1111/j.1365–2656.2008.01463.x.

Zuckerberg, B., A.M. Woods, and W.F. Porter. 2009. Poleward shifts in breeding bird distributions in New York State. Global Change Biology 15:1866–1883. doi:10.1111/j.1365–2486.2009.01878.x.

Index